AMD-Vol. 235

THICK COMPOSITES FOR LOAD BEARING STRUCTURES

presented at

THE 1999 ASME INTERNATIONAL MECHANICAL ENGINEERING CONGRESS AND EXPOSITION
NOVEMBER 14–19, 1999
NASHVILLE, TENNESSEE

sponsored by

THE APPLIED MECHANICS DIVISION, ASME

edited by

YAPA D. S. RAJAPAKSE
OFFICE OF NAVAL RESEARCH

GEORGE A. KARDOMATEAS
GEORGIA INSTITUTE OF TECHNOLOGY

THE AMERICAN SOCIETY OF MECHANICAL ENGINEERS
Three Park Avenue / New York, N.Y. 10016

ISBN No. 0-7918-1652-4

ISBN No. 99-76076

FOREWORD

Composite materials are supporting technological advancements in an increasing number of fields. Examples of recent active utilization of composites include transportation systems, marine vehicles, offshore structures, space vehicles, and civil infrastructure. Although thin plate construction was the thrust of most early applications, many current composite designs in these fields involve configurations classified as moderately thick structures to fully three-dimensional forms. Thus, there is active research in the mechanics community in many aspects of thick composites, including failure under monotonic or cyclic loading, three-dimensional effects including delamination, dynamic behavior, effects of manufacturing flaws such as fiber waviness, buckling and postbuckling behavior, and integrity of composite sandwich structures.

This volume contains twelve technical papers presented in a Symposium on Thick Composites for Load Bearing Structures, held at the 1999 International Mechanical Engineering Congress & Exposition (IMECE 99), in Nashville, Tennessee, November 14-19, 1999. It includes papers by some of the leading researchers in composite materials, and it represents a cross section of recent advances in the aforementioned topics of polymeric matrix composites research. The scope of these contributions spans a wide range of activities. Topics dealing with recent advances in laminated plates and shells include analysis of singularities in thick composites, mixed mode stress intensities at interface cracks, nonlinear behavior with fiber waviness, three-dimensional effects in failure response, shear deformation effects in the buckling and postbuckling behavior, dynamic behavior under multiaxial compression, subsonic and intersonic crack growth, and implosion behavior in composite shells under pressure. Recent advances in sandwich composites research dealing with stiffness and strength properties, failure modes including face-sheet/core delaminations, and high strain rate response, are also included in this volume.

The editors wish to thank all the authors for their excellent contributions to this volume. Collective efforts of this type by the research community are an important ingredient for technological progress.

Yapa D.S. Rajapakse
Office of Naval Research

George A. Kardomateas
Georgia Institute of Technology

CONTENTS

FAILURE OF A FIBER COMPOSITE LAMINA UNDER THREE-DIMENSIONAL STRESSES

Steven J. DeTeresa
University of California
Lawrence Livermore National Laboratory
Mechanical Engineering Department
7000 East Ave., MS L-342
Livermore, CA 94550
(925) 422-6466, (925) 422-2438 (FAX), deteresa1@llnl.gov

ABSTRACT

The efficient use of thick-section fiber composites requires a proven three-dimensional failure model. Numerous failure criteria have been proposed, but the lack of critical experimental results makes it difficult to assess the accuracy of these models. It is shown that the various predictions for failure of a lamina due to the simple state of uniaxial stress plus superposed hydrostatic pressure are disparate. These differences are sufficient to allow evaluation of failure criteria using data that has the normal scatter found for composite materials. A high-pressure test system for fiber composites is described and results for the effects of pressure on the transverse and longitudinal compression strengths of a carbon fiber/epoxy lamina are discussed. Results are compared with a few representative failure models.

INTRODUCTION

An outstanding problem in the analysis of fiber composite materials is the selection of a valid failure model. The problem isn't due to a lack of models—many have been proposed and reviewed in surveys (Rowlands, 1985, Nahas, 1986, Thom, 1998). Rather, the issue is the general acceptance of a single model as the most successful in capturing the response of fiber composites to multiaxial stresses. There is no single fiber composite model that has enjoyed proven success and therefore wide acceptance, as, for example, the von Mises criterion has in the case of isotropic materials.

A popular method of analyzing failure of laminated composite structures is to apply failure criteria to an individual layer of unidirectional material, the lamina, and then determine the effects of damage to the individual lamina on the eventual catastrophic failure of the structure. This approach has the potential benefit of providing failure predictions for any laminated structure using only the properties of the lamina. The recognition of this benefit has led to a great effort in applying lamina-based failure models. However, despite the extensive amount of

This work was performed under the auspices of the U. S. Department of Energy by the Lawrence Livermore National Laboratory under contract No. W-7405-Eng-48.

both theoretical and experimental work applied to this problem, there is still uncertainty regarding the effectiveness of this approach even when applied to simple plane stress states. A recent exercise to evaluate two-dimensional failure models has been organized by Hinton and Soden (1998). The approach is to compare predictions from several models to data obtained from a consistent set of experiments. While the use of critical experiments is necessary to evaluate failure models, thick-section composites require models and experiments that apply to fully three-dimensional stress states. To this end, we have begun to study the failure of fiber composites under simple triaxial stress states as a method to evaluate failure criteria. This paper summarizes an initial study of the effects of hydrostatic pressure on the uniaxial compression strengths of a fiber composite lamina.

BACKGROUND

The predictions of the proposed failure models for the simple stress state of pressure plus a uniaxial stress differ significantly and thus provide a means for their evaluation in spite of the normal scatter of test results for fiber composite materials. Much of the previous work to examine the effects of pressure on the strengths of fiber composites has been summarized by Hoppel et al. (1995). The work cited in this review and in a recent study by Zinoviev and Tsvetkov (1998) indicates that strength and failure modes of polymer-based fiber composites exhibit some dependence on pressure. However, the results varied with material, stress state and test method. Many of these studies resorted to nonstandard test specimens in order to perform tests in the confined volume of a pressure cell. Additionally, very few have reported results for a statistically significant number of tests at each pressure. Consequently, it is difficult to use these results to assess the accuracy of the reported effects of pressure on strength and then make the comparison with proposed failure models. In some cases it was observed that the failure mode was changed by the application of hydrostatic pressure (Zinoviev and Tsvetkov, 1998, Parry and Wronski, 1982). For the work reported by Parry and Wronski, (1982) the change in failure mode coincided with a change in the slope of strength (defined here as the stress difference) versus pressure. Since the effect of hydrostatic pressure on the uniaxial strengths of a lamina is a useful method for examining failure models, further results using more standard test methods and more widely studied materials are needed.

EXPERIMENTAL

All tests were conducted with the carbon fiber/epoxy composite material IM7/8551-7 (Hexcel, Inc. Dublin, CA). This material is useful for failure model evaluation because a full set of lamina elastic constants and strength data are available from several published sources (Jiang and Tennyson, 1989, Kallas and Hahn, 1990, Colvin and Swanson, 1990, Hercules, Inc., 1985). This data set is given in Table 1. Unidirectional 16-ply laminates were autoclave-cured following manufacturer's recommended procedures. All panels were inspected using ultrasonic C-scans and were found to be free from gross defects. The average cured ply thickness was 0.147 mm, which corresponds to a fiber volume fraction of roughly 58%.

Table 1. Elastic constants and strength data for IM7/8551-7.

Property	Value
Longitudinal Young's Modulus, E_{11} (GPa):	162.0
Transverse Young's Modulus, E_{22} (GPa):	8.34
Longitudinal Shear Modulus, G_{12} (GPa):	5.6
Transverse Shear Modulus, G_{23} (GPa):	—
Longitudinal Poisson's ratio, ν_{12}:	0.30
Transverse Poisson's ratio, ν_{23}:	0.50
Longitudinal Tensile Strength, X_T (GPa):	2.75
Longitudinal Compression Strength, X_C (GPa):	1.60
Transverse Tensile Strength, Y_T (MPa):	75
Transverse Compression Strength, Y_C (MPa):	185
Longitudinal Shear Strength, S_L (MPa):	80
F_{12} (MPa^{-2}):	8.412E-07
F_{23} (MPa^{-2}):	-3.914E-05, -2.678E-05, 3.914E-05

In order to compare our results for pressure effects with the data collected at ambient conditions, we sought to use standard test methods and specimen configurations. In some cases it was necessary to make modifications to these methods in order to adapt them to testing within a high-pressure cell. However, no method was deemed acceptable until it provided strengths at ambient pressure that were comparable to reported values. A carbon fiber/epoxy composite material was tested for the effects of pressure on both transverse and longitudinal compression strengths. For transverse tests, untabbed rectangular specimens were tested. Both uniform and reduced-width gage section ("dogbone") specimens were examined. The longitudinal strengths were measured using specimens having adhesive-bonded tabs.

Both types of specimens were tested using a two-piece cylindrical fixture that applied both end and shear loading. The design of the fixture was based on studies by Coguill and Adams (1999) and Haberle and Matthews (1994), who showed that combined loading as well as tab compression gave superior results. Set screws were used to apply compression to the tabs via a metal plate. Belleville washers were used under the set screws to maintain a high percentage of this compression when the specimen shrank in thickness under hydrostatic pressure. A precision brass tube was used to maintain the alignment between the two cylindrical fixtures and to prevent bending of the specimen.

The pressure system is comprised of a 1 GPa pressure cell connected at the bottom end to a servo-controlled high-pressure supply system and at the other end to a movable loading piston that contains an internal load cell. The internal load cell is a full-bridge strain gage that measures the differential axial load. The entire system is mounted in a servo-hydraulic test machine that controls the insertion of the piston and compression of the specimen in the pressure cell. Pressure was applied using kerosene as the medium. To test the sensitivity of the composite to kerosene at high pressure, specimens were tested for transverse compression strength at ambient conditions before and after exposure to 345 MPa pressure for 30 minutes. This exposure test was considered severe since the actual time at pressure during a typical test for strength was less than 15 minutes. Results for compression strength and weight change before and after showed that the composite was unaffected by exposure to high-pressure kerosene. Consequently, all tests were conducted with bare specimens.

Compression tests were conducted at 0, 172, and 345 MPa pressure with at least five repeats at each condition.

RESULTS

Failure Models

For comparison with data, only a few representative failure models were considered. As additional experimental results are collected in the future, a more extensive comparison will be made. Here results are compared with the maximum stress and strain theories, the Tsai-Wu tensor polynomial (Tsai and Wu, 1971), the Hashin criterion (Hashin, 1980), and the Christensen stress-based criterion (Christensen, 1998). The last two give separate failure criteria for matrix- and for fiber-dominated failures. The Hashin model requires a value for the transverse shear strength, but due to experimental difficulties, this value is normally not measured. Instead, the matrix failure criterion of the Christensen theory can be used to calculate the transverse shear strength from the transverse tensile and compression strengths. When this calculated value of transverse shear strength is used in the matrix compression failure criterion of Hashin's theory, it becomes identical to the Christensen matrix failure criterion.

Failure Due to Hydrostatic Pressure

Prior to measurement of the effect of pressure on uniaxial strength, the composite was examined for evidence of failure under pure hydrostatic pressure. Although it is commonly accepted that homogeneous, isotropic materials are not susceptible to failure under moderate levels of hydrostatic pressure, yielding of unidirectional organic fiber and metal matrix composites has been observed (Zinoviev and Tsvetkov, 1998, Wakashima and Umekawa, 1976). In both cases, exposure to pressures below 690 MPa resulted in negative plastic strain (shrinkage) transverse to the fiber and positive plastic strain (elongation) in the fiber direction. Due to the mismatch in elastic properties between fiber and matrix, hydrostatic pressure can induce deviatoric stresses in both components and initiate yielding.

A simple square array of circular cross-section fibers was modeled to estimate the magnitude of these stresses. Under hydrostatic pressure it was found that the maximum stress difference occurs in the region where fibers are closest to each other. For elastic constituents, the stress difference was found to be $P/2$, where P is the externally applied hydrostatic pressure. Typical epoxy matrix materials exhibit yield at shear stresses in the range of 50 to 70 MPa. Therefore, it is possible that pure pressure in the range of 200–280 ksi could initiate local yielding of the matrix in a unidirectional composite.

Some of the failure models for fiber composites predict failure under pure hydrostatic pressure. Obviously, limit failure models such as the popular maximum stress and strain criteria reach failure conditions when pressure or the strain due to pressure equals the limit in any direction. For most fiber composite lamina, this will occur in the transverse direction first. The tensor polynomial failure models also reach a failure condition under pure pressure. For example, the Tsai-Wu model predicts failure at the critical pressure

$$P_{cr} = \frac{-(F_1+2F_2)-\sqrt{(F_1+2F_2)^2+4(F_{11}+2F_{22}+4F_{12}+2F_{23})}}{2(F_{11}+2F_{22}+4F_{12}+2F_{23})} \tag{1}$$

With the values given in Table 1 for the interactive term F_{23}, the critical pressure for failure of IM7/8551-7 ranges from about 110 to 250 MPa, which is less than the maximum pressure used in the studies conducted here. The previously discussed result for the effects of pressure on the retention of transverse compression strength at ambient conditions indicates that there is no evidence of failure. Measurement of specimen dimensions before and after exposure to 345 MPa pressure showed no change to within a measurement error of 0.01% strain. Despite the potential for yielding of a lamina under pure pressure and the predictions by some failure models for failure under this loading, the experimental evidence for this particular composite material indicates that its response is elastic under hydrostatic pressure up to 345 MPa.

Transverse Compression Strength

To compare the effect of pressure with failure models, it was decided to define strength as the stress difference for the loading direction rather than the total stress. Thus for the 1-direction aligned with the fiber, uniaxial compression under pressure is given by

$$\sigma_{11} = P, \sigma_{22} = P + \sigma_{22}^{*}, \sigma_{33} = P \tag{2}$$

where σ_{22}* is the stress difference or transverse strength of the composite. The corresponding case for longitudinal compression is obvious. Here both stresses and pressure are considered positive for tension. With this definition of strength, pressure-independent (Mises-like) behavior is obvious.

The results for testing at the different pressure levels are summarized in Table 2, showing that transverse compression strength increases with pressure. Although there is good agreement between the strengths obtained using the standard uniform-width and dogbone specimens, the scatter is less for the latter. The values for strength at ambient pressure are identical to reported values, indicating that the test method is valid. The uniform-width specimens sometimes failed at the fixture while the dogbone specimens always failed in the reduced width gage section away from the fixture. For this reason, the strength data from the dogbone specimens were considered more accurate. At all three pressures, specimens failed by shear through the thickness. No change in failure mode with pressure was observed.

Table 2. Effect of pressure on the transverse compression strength of IM7/8551-7.

	Compression Strength (MPa)	
Pressure (MPa)	**Straight Specimen**	**Dogbone Specimen**
0	173 (4.9)	186 (3.0)
172	238 (7.6)	237 (7.6)
345	283 (17.2)	272 (10.3)

Values in parentheses are standard deviations.

The comparison of these results with the failure criteria is shown in Fig. 1. The Hashin and Christensen models give the correct trend, but overpredict the magnitude of strength increase with pressure. The limit and tensor polynomial failure models predict the opposite trend as well as failure due to pure pressure at values less than those attained in the experiments.

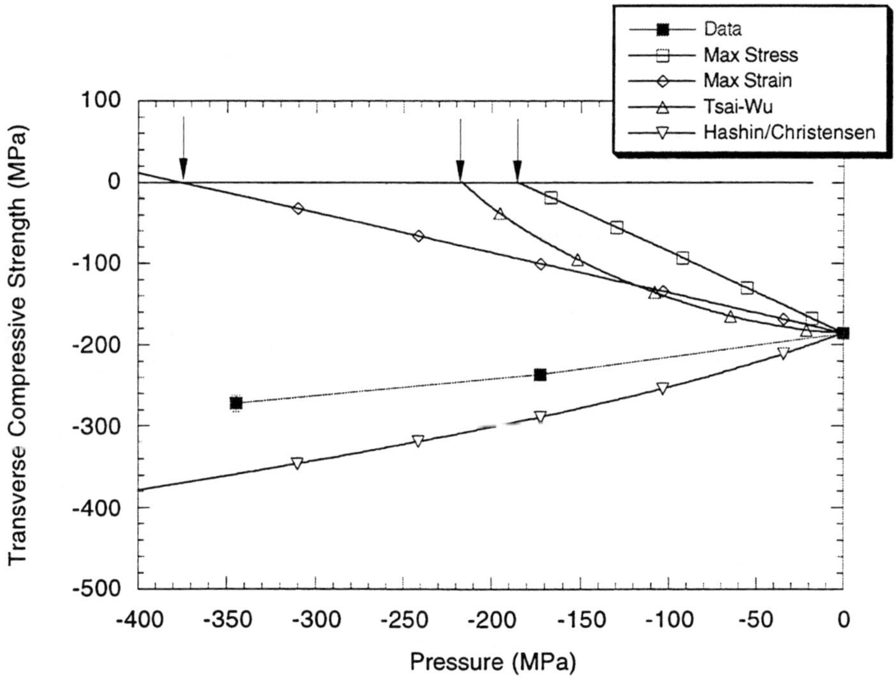

Figure 1. Experimental results and predictions for the effect of pressure on transverse compression strength of IM7/8551-7.

Longitudinal compression strength

For compression in the fiber direction, the stress state is given by

$$\sigma_{11} = P + \sigma_{11}^{*}, \sigma_{22} = P, \sigma_{33} = P \tag{3}$$

where σ_{11}* is the stress difference or longitudinal strength of the composite. The initial strength at ambient pressure was 1.46 GPa, which is slightly lower than reported values. This difference could be attributed to fiber volume fraction, which was somewhat lower in the samples used here.

In all cases, failure occurred in the gage section and usually at the end of a tab, which is typical for this type of specimen. No significant change in failure mode was observed at the higher pressures. Pressure resulted in a decrease in strength at a rate that was close to predictions based on the maximum strain and Christensen theories as shown in Fig. 2. The maximum stress and Hashin theories, which are identical for this stress state, predict a much greater loss in strength with pressure. The tensor polynomial theory predicts a significant increase in strength at low pressures followed by a large decrease in strength. It is interesting to compare these results with those obtained by Parry and Wronski (1982). They observed an initial decrease in strength followed by a strength increase at the point the failure mode changed. This reversal in pressure-dependence is predicted only by the Christensen theory, but the predicted effect is much less than found by Parry and Wronski. In fact, these authors state that for their composite the tensile and compression strengths were equivalent. According to the Christensen theory, equivalent strengths yield Mises-like behavior. In other words, strength is independent of pressure.

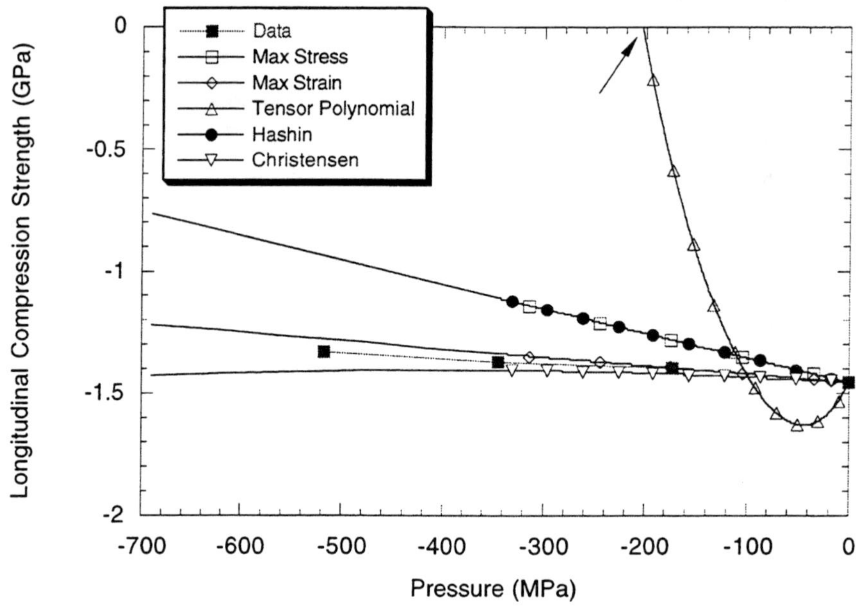

Figure 2. Experimental results and predictions for the effect of pressure on the longitudinal compression strength of IM7/8551-7.

SUMMARY AND CONCLUSIONS

The effect of pressure on the transverse and longitudinal compression strengths of a carbon fiber/epoxy lamina was measured using newly developed techniques for high-pressure testing of composites. Transverse strength was found to increase and longitudinal strength to decrease with pressure. Unlike previous studies, no change in failure mode due to superposed pressure was observed. The effect of pressure on strength was qualitatively predicted by the Hashin and Christensen models for transverse compression failure and by the maximum strain and Christensen models for longitudinal compression failure. These results and those of future studies will be used to critically evaluate proposed three-dimensional failure criteria for thick-section composites.

ACKNOWLEDGEMENTS

The author gratefully acknowledges the support by the Office of Naval Research under Contract No. N00014-96-F-0422, technical monitor Dr. Y. Rajapakse, and by the Joint DOE/DoD Technology Development Munitions Program. The contributions of Mr. Roberto Sanchez and Mr. Greg Larsen to perform the high-pressure tests and of Mr. Will Andrade and Mr. Jeff Petersen to develop the machining techniques and prepare the specimens are also greatly appreciated.

REFERENCES

Christensen, R. M., "The numbers of elastic properties and failure parameters for fiber composites," *J. Eng. Mater. Tech. Trans. ASME*, **120**, 1998, pp. 110–113.

Coguill, S. L. and Adams, D. F., "Use of the Wyoming combined loading compression fixture to test unidirectional composites," *Proc. 44th International SAMPE Symp.*, May, 1999.

Colvin, G. E. and Swanson, S. R., "Mechanical characterization of IM7/8551-7 carbon/epoxy under biaxial stress," *J. Eng. Mater. Tech., Trans. ASME*, **112**(1), 1990, pp. 61–67.

Haberle, J. G. and Matthews, F. L., "An improved technique for compression testing of unidirectional fibre-reinforced plastics; development and results," *Composites*, **25**(5), 1994, pp. 358–371.

Hashin, Z., "Failure criteria for unidirectional fiber composites," *J. Appl. Mech.*, **47**, 1980, pp. 329–334.

Hercules, Inc., "Magnamite 8551-7 Tough Resin, Graphite Prepreg Tape and Fabric Data Summary," Hercules Aerospace Products Group, Technical Bulletin, November, 1985.

Hinton, M. J. and Soden, P. D., "Predicting failure in composite laminates: the background to the exercise," *Comp. Sci. Tech.*, **58**(7), 1998, 1001.

Hoppel, C. P. R., Bogetti, T. A., and Gillespie, J. W. Jr., "Effects of Hydrostatic Pressure on the Mechanical Behavior of Composite Materials," *J. Thermoplastic Comp. Mat.*, **8**(4),1995, pp. 375–409.

Jiang, Z. and Tennyson, R. C., *J. Comp. Mater.*, **23**(3), 1989, pp. 208–231.

Kallas, M. N. and Hahn, H. T., "A failure criterion for composites under 3-dimensional state of stress," *Proc. 35th International SAMPE Symp.*, April 2–5, 1990, pp. 674–686.

Nahas, M. N., "Survey of failure and post-failure theories of laminated fiber-reinforced composites," *J. Comp. Tech. Res.*, **8**(4) 1986, pp. 138–153.

Parry, T. V. and Wronski, A. S., "Kinking and compressive failure in uniaxially aligned carbon fibre composite tested under superposed hydrostatic pressure," *J. Mater. Sci.*, **17**, 1982, pp. 893–900.

Rowlands, R. E., "Strength (failure) theories and their experimental correlation," in *Failure Mechanics of Composites*, G. C. Sih and A. M. Skudra, eds., North-Holland, Amsterdam, 1985, pp. 71–125.

Thom, H., "A review of the biaxial strength of fibre-reinforced plastics," *Comp. Pt A.*, **29A**, 1998 pp. 869–886.

Tsai, S. W. and Wu, E. M., "A general theory of strength for anisotropic materials," *J. Comp. Mater.* **5**, 1971, p. 58.

Wakashima, K. and Umekawa, S., "Fiber-composite response to hydrostatic stress," *Metall. Trans.*, **7A**, 1976, pp. 1952–1954.

Zinoviev, P. A. and Tsvetkov, S. V., "Mechanical properties of unidirectional organic-fiber-reinforced plastics under hydrostatic pressure," *Comp. Sci. Tech.* **58**, 1998, pp. 31–39.

AMD-Vol. 235, Thick Composites for Load Bearing Structures
ASME 1999

ANALYSIS OF NONLINEAR FLEXURAL BEHAVIOR OF THICK COMPOSITES WITH FIBER WAVINESS

H. -J. Chun
School of Electrical and Mechanical Engineering, Yonsei University

S. W. Lee
Department of Mechanical Engineering, Yonsei University,
134 Shinchon-Dong, Seodaemun-Ku, Seoul, 120-749, Korea

I. M. Daniel
Departments of Mechanical and Civil Engineering, Northwestern University,
Evanston, IL 60208, USA

Abstract

A finite element analysis model was developed to predict flexural behavior of thick composites with uniform, graded and localized fiber waviness. In the analyses, material and geometrical nonlinearties due to fiber waviness were incorporated into the model utilizing energy density and an incremental method. In the model, two kinds of geometrical nonlinearity were considered, one due to reorientation of fibers and the other due to difference of curvatures from one finite element to another during deformation. The finite element analyses utilize the iterative mapping method to incorporate these geometrical nonlinear factors. The model was used to predict not only the flexural behavior of a flat thick composite plate but also of a thick composite plate with initial curvature. Flat composite specimens with various degrees of fiber waviness were fabricated and four-point flexural tests were conducted. The predicted nonlinear behavior by the current model was compared with results from the thin slice model [7] and experiments. Good agreement was observed among them.

INTRODUCTION

Fiber waviness is one of the manufacturing defects frequently encountered in thick composite structures. Its characteristics can be represented by the through the thickness undulation of the fibers. It appears frequently during the filament winding process of fabricating thick cylindrical composite structures. It is, therefore, very important to study effects of fiber waviness on initially curved composite structures.

A number of studies have been conducted on the behavior of thick composites with fiber waviness. Shuart [1] considered both in-plane and out-of–plane waviness in a microbuckling model for multidirectional laminates. Telegadas and Hyer [2] conducted finite element analyses on composites with fiber waviness. Chou and Takahashi [3] predicted the tensile stress-strain response of flexible wavy fiber composites. Hsiao and Daniel [4-6] investigated the effect of fiber waviness on compressive behavior of thick composites. Chun *et al.* [7] investigated nonlinear behavior of thick composites with fiber waviness using an analytical model(thin slice model). Kedward *et al.* [8] investigated flexural behavior of simply curved composite shapes utilizing a finite element method. Ganapathi *et al.* [9] developed a membrane-share-bending element for geometrically nonlinear behavior of laminates. However, none of these studies investigated the nonlinear flexural behavior of thick composites considering both effects of fiber waviness and initial curvature.

In this study, the flexural behavior of thick curved composite beams with various initial curvatures is investigated theoretically and experimentally. The effects of material and geometric nonlinearities are also incorporated into the finite element analysis by employing the strain density and iterative mapping method.

ANALYSIS

Unidirectional composites examined are composed of continuous fibers with sinusoidal waviness in a matrix. Due to material as well as geometric factors, the flexural behavior of these composites is generally nonlinear under finite deformation. A FEA(finite element analysis) model is proposed for predicting the nonlinear behavior of composites with fiber waviness.

Uniform Fiber Waviness Model :

1) Flat Beam

Figure 1 illustrates a representative volume of a flat uniform fiber waviness model. It is assumed that all fibers are parallel to each other and have sinusoidal curvature along a spatial coordinate direction (*x*-axis). The representative volume is then divided into subelements in both *x* and *z* directions. Then, each subelement can be treated as an off-axis unidirectional laminate.

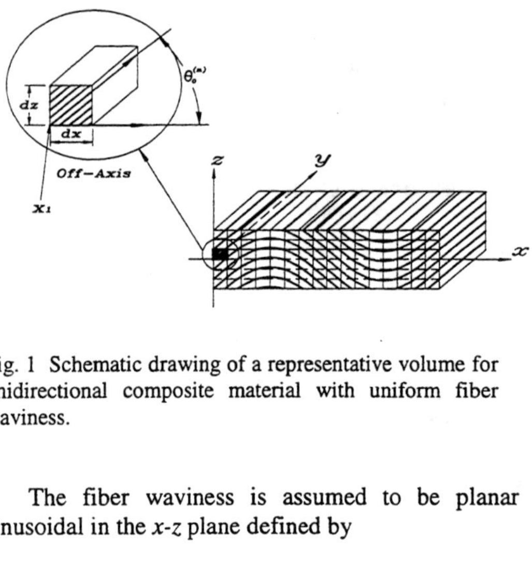

Fig. 1 Schematic drawing of a representative volume for unidirectional composite material with uniform fiber waviness.

The fiber waviness is assumed to be planar sinusoidal in the *x*-*z* plane defined by

$$z = a\sin\frac{2\pi x}{\lambda} \tag{1}$$

It is necessary to relate the initial fiber orientation, $\theta_0^{(n)}$, to the wave parameters a and λ which are the amplitude and wavelength of the wavy fiber, respectively. The fiber orientation, θ, at $x = x_1$ is related to x and z and obtained as follows

$$\tan\theta\Big|_{x_1} = \frac{dz}{dx} = \frac{2\pi a}{\lambda}\cos\frac{2\pi x_1}{\lambda} \tag{2}$$

Therefore, from Fig. 1 and equation (2), the initial fiber orientation of the nth subelement , $\theta_0^{(n)}$, can be written as

$$\theta_0^{(n)} = \frac{1}{2}\left[\tan^{-1}\left(\frac{dz}{dx}\right)_{x=x_1} + \tan^{-1}\left(\frac{dz}{dx}\right)_{x=x_1+dx}\right]$$

$$= \frac{1}{2}\left[\tan^{-1}\left(\frac{2\pi a}{\lambda}\cos\frac{2\pi x_1}{\lambda}\right) + \tan^{-1}\left(\frac{2\pi a}{\lambda}\cos\frac{2\pi(x_1+dx)}{\lambda}\right)\right] \tag{3}$$

Energy density (W) is used to incorporate the material nonlinearity in the model. For nonlinear elastic composites, the fourth order expansion of W is considered

$$\begin{aligned}
W &= W\left(\varepsilon_{11}, \varepsilon_{22}, \varepsilon_{33}, \gamma_{12}^2, \gamma_{23}^2, \gamma_{13}^2, \gamma_{12}\gamma_{23}\gamma_{13}\right) \\
&= \frac{1}{2}C_{11}\varepsilon_{11}^2 + \frac{1}{2}C_{22}\varepsilon_{22}^2 + \frac{1}{2}C_{33}\varepsilon_{33}^2 \\
&+ \frac{1}{2}C_{44}\gamma_{23}^2 + \frac{1}{2}C_{55}\gamma_{13}^2 + \frac{1}{2}C_{66}\gamma_{12}^2 \\
&+ C_{12}\varepsilon_{11}\varepsilon_{22} + C_{13}\varepsilon_{11}\varepsilon_{33} + C_{23}\varepsilon_{22}\varepsilon_{33} \\
&+ \frac{1}{3}C_{111}\varepsilon_{11}^3 + \frac{1}{3}C_{222}\varepsilon_{22}^3 + \frac{1}{3}C_{333}\varepsilon_{33}^3 \\
&+ \frac{1}{4}C_{1111}\varepsilon_{11}^4 + \frac{1}{4}C_{2222}\varepsilon_{22}^4 + \frac{1}{4}C_{3333}\varepsilon_{33}^4 \\
&+ \frac{1}{4}C_{4444}\gamma_{23}^4 + \frac{1}{4}C_{5555}\gamma_{13}^4 + \frac{1}{4}C_{6666}\gamma_{12}^4
\end{aligned} \tag{4}$$

where $\varepsilon_{ii}, \gamma_{ij}$ and C's are on-axis normal strains, on-axis shear strains and stiffnesses, respectively.

The stress-strain relations can be derived from the energy density as follows

$$\sigma_{ij} = \frac{\partial W}{\partial \varepsilon_{ij}} \tag{5}$$

If the nonlinear coupling terms between normal and shear strains are neglected, the stress-strain relations referred to material coordinates can be expressed in following matrix form

$$[\sigma]_{1,2,3} = [C^*][\varepsilon]_{1,2,3} \tag{6}$$

where the corresponding stiffnesses are expressed as follows

$$
\begin{aligned}
C_{11}^* &= C_{11} + C_{111}\varepsilon_{11} + C_{1111}\varepsilon_{11}^2 \\
C_{22}^* &= C_{22} + C_{222}\varepsilon_{22} + C_{2222}\varepsilon_{22}^2 \\
C_{33}^* &= C_{33} + C_{333}\varepsilon_{33} + C_{3333}\varepsilon_{33}^2 \\
C_{44}^* &= C_{44} + C_{4444}\gamma_{23}^2 \\
C_{55}^* &= C_{55} + C_{5555}\gamma_{23}^2 \\
C_{66}^* &= C_{66} + C_{6666}\gamma_{12}^2 \\
C_{12}^* &= C_{12}, C_{13}^* = C_{13}, C_{23}^* = C_{23}
\end{aligned} \tag{7}
$$

For the plane problem, the relations between on-axis strains and off-axis strains are obtained by using transformation relations as follows

$$
\begin{bmatrix} \varepsilon_{11} \\ \varepsilon_{33} \\ \gamma_{13} \end{bmatrix} = \begin{bmatrix} m^2 & n^2 & mn \\ n^2 & m^2 & -mn \\ -2mn & 2mn & m^2 - n^2 \end{bmatrix} \begin{bmatrix} \varepsilon_x \\ \varepsilon_z \\ \gamma_{xz} \end{bmatrix} \tag{8}
$$

where

$\varepsilon_{22} = \varepsilon_y, \gamma_{12} = \gamma_{23} = 0$

$m = \cos\theta^{(n)}, \quad n = \sin\theta^{(n)}$

$\theta^{(n)}$ = fiber orientation in the nth subelement after deformation

The on-axis stiffness can be obtained from the stress-strain relation as

$$
[\sigma]_{1,3} = [Q]_{1,3}[\varepsilon]_{1,3} \tag{9}
$$

$$
[Q]_{1,3} = \begin{bmatrix} Q_{11} & Q_{13} & 0 \\ Q_{13} & Q_{33} & 0 \\ 0 & 0 & Q_{55} \end{bmatrix}
$$

where

$$
Q_{11} = C_{11}^* - \frac{C_{12}^* C_{12}^*}{C_{22}^*}, \quad Q_{13} = C_{13}^* - \frac{C_{13}^* C_{13}^*}{C_{22}^*},
$$

$$
Q_{33} = C_{33}^* - \frac{C_{23}^* C_{23}^*}{C_{22}^*}, \quad Q_{55} = C_{55}^*
$$

Then, the off-axis stiffness matrix can be transformed as

$$
[Q^*] = [T^{-1}][Q][T]
$$

$$
= \begin{bmatrix} Q_{xx} & Q_{xz} & 2Q_{xs} \\ Q_{zx} & Q_{zz} & 2Q_{zs} \\ Q_{sx} & Q_{sz} & 2Q_{ss} \end{bmatrix} \tag{10}
$$

For a particular subelement, the changed fiber orientation due to the deformation is obtained as a function of strains, coordinates of nodes and nodal displacements. Figure 2 illustrates the deformation and associated nodal displacements of a subelement.

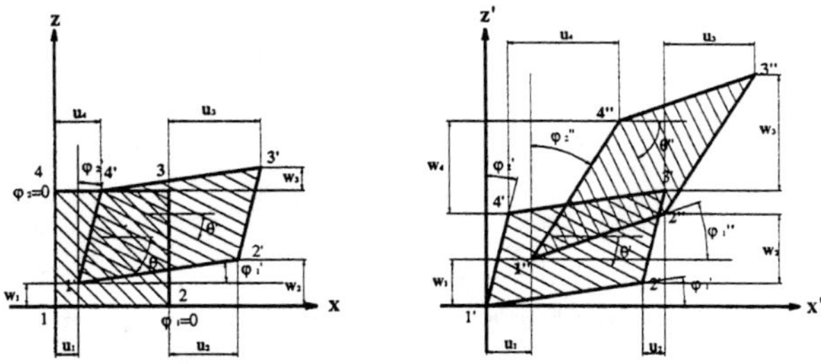

Fig. 2 Schematic drawing showing the deformation of subelement.

From figure 2 and equation (3), the fiber orientation of a deformed subelement, $\theta^{(n)}$, is given as

$$
\theta^{(n)} = \tan^{-1}\left[\Delta\varphi_1 - \Delta\varphi_2 + \frac{1+\varepsilon_z^{(n)}}{1+\varepsilon_x^{(n)}} \tan\theta_0^{(n)}\right] \tag{11}
$$

where

$$
\varphi_1 = \tan^{-1}\frac{z_2' - z_1'}{x_2' - x_1'}, \quad \varphi_2 = \tan^{-1}\frac{x_3' - x_1'}{z_3' - z_1'}
$$

$$
\varepsilon_x^{(n)} = \frac{\partial u}{\partial x} = \frac{1}{2}\left(\frac{u_2 - u_1}{x_2 - x_1} + \frac{u_3 - u_4}{x_3 - x_4}\right)
$$

$$
\varepsilon_z^{(n)} = \frac{\partial w}{\partial z} = \frac{1}{2}\left(\frac{w_4 - w_1}{z_4 - z_1} + \frac{w_3 - w_2}{z_3 - z_2}\right)
$$

$$
\gamma_{xz}^{(n)} = \frac{\partial w}{\partial x} + \frac{\partial u}{\partial z}
= \frac{1}{2}\left(\frac{w_2 - w_1}{x_2 - x_1} + \frac{w_3 - w_4}{x_3 - x_4} + \frac{u_4 - u_1}{z_4 - z_1} + \frac{u_3 - u_2}{z_3 - z_2}\right)
$$

$\varepsilon_x^{(n)}$ and $\varepsilon_z^{(n)}$ are strains in the x and z directions for the nth element, respectively; u_i, w_i, x_i and z_i are displacements and coordinates in the x and z directions for the ith node, respectively.

From the equilibrium equation for finite element analysis, the relation between force vector $[F]$ and displacement vector $[d]$ is expressed as follows

$$[K][d]=[F]$$

$$[K_{ij}]^e = \int_{\Omega^e} [B_i]^T[D][B_j]d\Omega^e = \int_{-1}^{1}\int_{-1}^{1} [B_i]^T[D][B_j]|J|d\xi d\eta \tag{12}$$

where

$[K]$ = global stiffness matrix

$[K_{ij}]^e$ = element stiffness matrix

$[B]$ = strain shape function matrix

$|J|$ = determinant of Jacobian matrix

ξ, η : element coordinates in two dimensions

$$[B_m] = \begin{bmatrix} \frac{\partial N_m}{\partial x} & 0 \\ 0 & \frac{\partial N_m}{\partial z} \\ \frac{\partial N_m}{\partial z} & \frac{\partial N_m}{\partial x} \end{bmatrix}$$

$[N]$ = shape function matrix

N_m = shape function

m = nodal number

$$N_1 = \frac{1}{4}(1-\xi)(1-\eta),\ N_2 = \frac{1}{4}(1+\xi)(1-\eta)$$

$$N_3 = \frac{1}{4}(1+\xi)(1+\eta),\ N_4 = \frac{1}{4}(1-\xi)(1+\eta)$$

Once the stiffness matrix is determined, for given boundary conditions and force vector $[F]$, the corresponding displacement vector $[d]$ is obtained by the Newton-Raphson method. The strain vector is determined from the displacement vector and strain-displacement relations. The stress vector is then derived from stress-strain relations. Iterative mapping is used to recalculate $[K]$ due to fiber orientation during incremental deformation. An incremental method is adopted to incorporate the nonlinear characteristics into the analysis. These procedures are repeated till the final value of load is reached.

2) Beam with Initial Curvature :

Figure 3 shows a schematic drawing for a uniform fiber waviness model with initial curvature. It is assumed in the model that the thick composite beam has uniform fiber waviness with initial curvature. The representative volume is then divided into subelements along the r-ϕ coordinates. Then, each subelement can be treated as an off-axis lamina similar to that of the flat uniform waviness model. Additional x'-z' coordinates have been introduced to facilitate the analysis. It is also assumed that the x'-z' coordinates share the same origin.

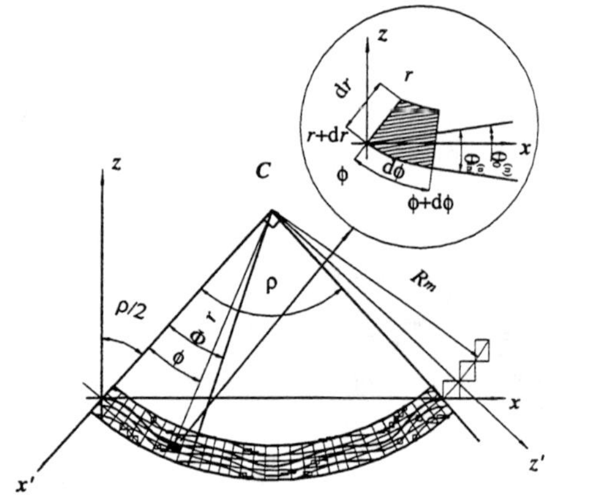

Fig. 3 Schematic drawing of a representative volume for unidirectional composite material with both uniform fiber waviness and initial curvature.

From Figure 3, the fiber waviness is assumed to be sinusoidal in the r-ϕ coordinates and expressed as

$$r = -a\sin\frac{2\pi\phi}{\Phi} + R_m \tag{13}$$

where a, Φ, R_m and ϕ are the amplitude of fiber waviness, the angle for a wavelength of fiber waviness, the initial radius of curvature of the midsurface and the angle to the nth subelement from one reference end of the representative volume, respectively.

The relations between x'-z' and r-ϕ coordinates are expressed as follows

$$x' = r\cos\phi,\ z' = r\sin\phi$$

$$r = \sqrt{x'^2 + z'^2},\ \phi = \tan^{-1}\frac{z'}{x'} \tag{14}$$

then, equation (13) can be rewritten as

$$r = \sqrt{x'^2 + z'^2} - a \sin \frac{2\pi \tan^{-1} \frac{z'}{x'}}{\Phi} + R_m \qquad (15)$$

Transformation relations between x'-z' and x-z coordinates are expressed as

$$\begin{aligned}
x &= -x'\sin\frac{\rho}{2} + z'\cos\frac{\rho}{2} + R_m \sin\frac{\rho}{2} \\
z &= -x'\cos\frac{\rho}{2} - z'\sin\frac{\rho}{2} + R_m \cos\frac{\rho}{2} \qquad (16) \\
x' &= -\left[x - R_m\cos\frac{\rho}{2}\right]\sin\frac{\rho}{2} - \left[z - R_m\cos\frac{\rho}{2}\right]\cos\frac{\rho}{2} \\
z' &= \left[x - R_m\cos\frac{\rho}{2}\right]\cos\frac{\rho}{2} - \left[z - R_m\cos\frac{\rho}{2}\right]\sin\frac{\rho}{2}
\end{aligned}$$

where ρ is the total angle of the beam under analysis. Then, the initial fiber orientation $\theta_{ini}^{(n)}$ referred to the $r-\phi$ coordinates is expressed as

$$\theta_{ini}^{(n)} = \frac{1}{2}\left[\tan^{-1}\left(\frac{1}{r}\frac{dr}{d\phi}\right)_{\phi} + \tan^{-1}\left(\frac{1}{r}\frac{dr}{d\phi}\right)_{\phi+d\phi}\right] \qquad (17)$$

Finally, the initial fiber orientation in x-z coordinates, $\theta_0^{(n)}$, is expressed as

$$\theta_0^{(n)} = \theta_{ini}^{(n)} - \sin^{-1}\left[\frac{1}{d\theta}\left(1 - \cos\left(\frac{\rho}{2} - \phi\right)\right)\right] \qquad (18)$$

Using equation (18), the flexural behavior of the uniform fiber waviness model with initial curvature is analyzed following similar procedures conducted for the flat uniform fiber waviness model.

Graded Fiber Waviness Model

1) Flat Beam

Figure 4 shows a representative volume of graded fiber waviness model. The representative volume consists of thicknesswise volume fraction (V_{ct}) of graded fiber waviness and volume fraction without fiber waviness. It is assumed that the amplitude of fiber waviness decays linearly from a maximum at the midsurface to zero at certain surfaces parallel to the outer surfaces.

The amplitude of the lth element along the z-axis is given as

$$a_l = a_0\left(1 - \frac{2|z_l|}{h \cdot V_{ct}}\right)\left(1 - u\left(|z_l| - \frac{h \cdot V_{ct}}{2}\right)\right) \qquad (19)$$

where a_0, z_l, h, u and $V_{ct} = h_w / h$ are the maximum amplitude of fiber waviness, distance from midplane to the lth subelement, thickness of the flat beam, unit step function and thicknesswise volume fraction ratio, respectively.

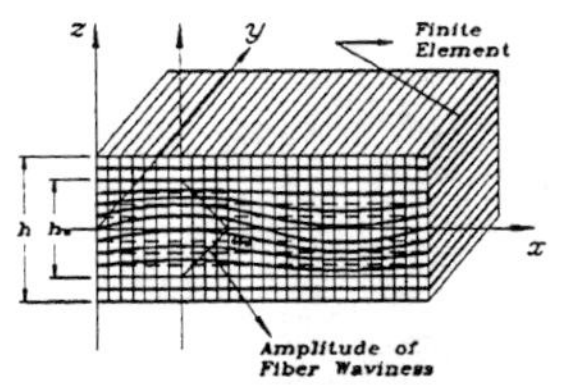

Fig. 4 Schematic drawing of a representative volume for flat beam with graded fiber waviness.

2) Beam with Initial Curvature

Figure 5 shows a schematic drawing for the graded fiber waviness model with initial curvature. The amplitude of the lth element along the r-axis is given as

$$a_l = a_0\left(1 - \frac{2|r_l|}{h \cdot V_{ct}}\right)\left(1 - u\left(|r_l| - \frac{h \cdot V_{ct}}{2}\right)\right) \qquad (20)$$

where a_0, z_l, h, u and $V_{ct} = h_w / h$ are the maximum amplitude of fiber waviness, distance from midsurface to the lth subelement, thickness of the beam with initial curvature, unit step function and thicknesswise volume fraction ratio, respectively.

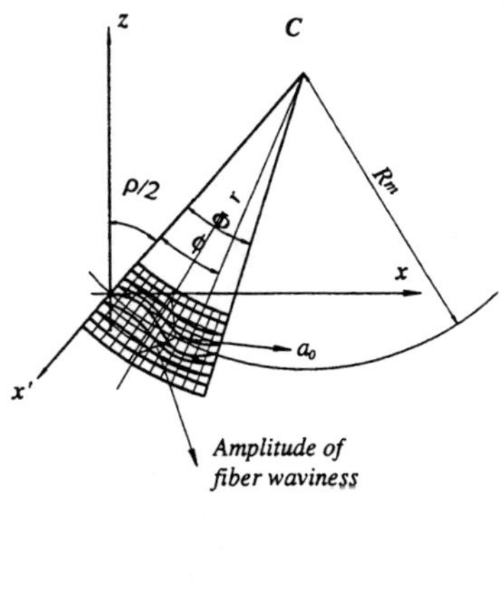

Fig. 5 Schematic drawing for graded fiber waviness model with initial curvature.

Localized Fiber waviness Model

1) Flat Beam

Figure 6 shows a representative volume containing a single period of localized fiber waviness. The representative volume consists of lengthwise volume fraction ($V_{cl} = L_w / L$) of localized fiber waviness and volume fraction without fiber waviness. h, h_w, L and L_w are the thickness of representative volume, thickness of volume containing fiber waviness, length of representative volume and length of volume containing fiber waviness, respectively.

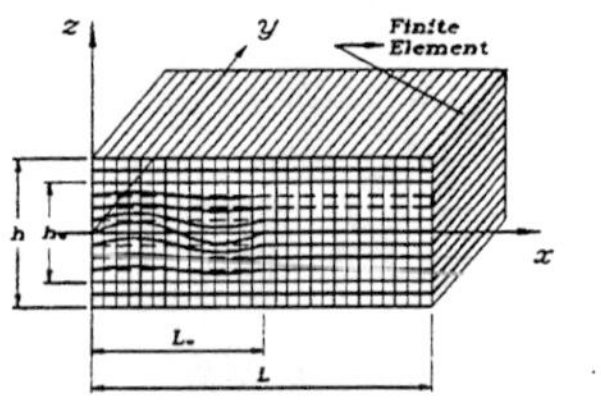

Fig. 6 Schematic drawing of a representative volume and coordinates for flat beam with localized fiber waviness.

2) Beam with Initial Curvature

The flexural behavior of the localized fiber waviness model with initial curvature is analyzed following similar procedures as those conducted for flat localized fiber waviness model in cylindrical coordinates.

Analyses of Beams with Initial Curvatures :

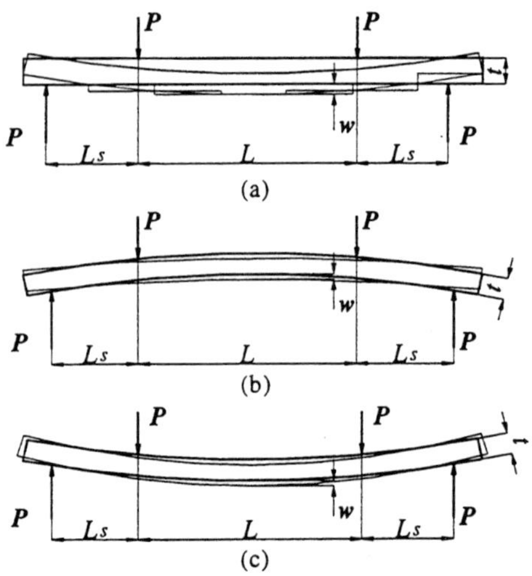

Fig. 7 Three types of beams, (a) flat beam, (b) convex beam, (c) concave beam.

The FEA model is applied to three types of thick composite beams with fiber waviness : flat, convex and concave beams. Figure 7 illustrates three types of beams with initial curvature under four-point flexural loading.

EXPERIMENTAL PROCEDURES

The material investigated in this study was DMS 2224 graphite/epoxy composite material (Hexcel Co. Inc.). To determine the elastic properties and the nonlinear stress-strain behavior of this material, quasi-static tensile tests with standard specimens were conducted in an MTS servo-hydraulic testing system. Also quasi-static compression tests with standard specimens were performed using the IITRI compression test fixture.

A fabrication technique was developed for producing thick composites with controlled fiber waviness. Molds with various sinusoidal waves were used to fabricate the composites with fiber waviness in an autoclave following a special two-step curing cycle. Specimens with three waviness ratios (amplitude/wavelength) were fabricated. The fiber waviness ratios were 0.011, 0.034 and 0.059. The specimens were 150mm long and 10mm wide. The test coupons were instrumented strain gages both on the top and bottom sides. Figure 8 shows the geometry of the four-point flexural coupons. Four-point flexural tests were conducted in a servo-hydraulic testing machine while monitoring the deflections, strains and applied loads. The deflections of the composite beam were measured with the home-made deflection meter.

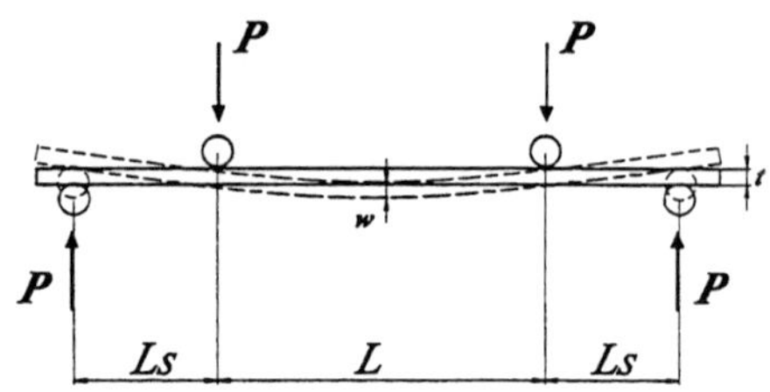

Fig. 8 Schematic drawing of four-point flexural test (L = 70mm, L_s = 35mm, t = 5.2mm).

RESULTS AND DISCUSSION

The mechanical properties and higher order stiffnesses for material nonlinearity were obtained from tensile and compressive stress-strain curves and are listed in Tables 1 and 2, respectively.

Table 1 Mechanical properties of DMS 2224 graphite/epoxy composite material

Properties	Values
Tensile Longitudinal Modulus E_{1t}, GPa	120
Tensile Transverse Modulus E_{2t}, GPa	9.07
Tensile In-Plane Shear Modulus G_{12t}, GPa	4.73
Tensile Major Poisson's Ratio ν_{12t}	0.286
Tensile Minor Poisson's Ratio ν_{21t}	0.022
Compressive Longitudinal Modulus E_{1c}, GPa	122
Compressive Transverse Modulus E_{2c}, GPa	9.11
Compressive Major Poisson's Ratio ν_{12c}	0.293
Compressive Minor Poisson's Ratio ν_{21c}	0.022

Table 2 Stiffnesses of DMS 2224 composite material

Stiffness (GPa)	Value	Stiffness (GPa)	Value
C_{11t}	122.45	C_{11c}	124.30
C_{111t}	522.2	C_{111c}	228.5
C_{1111t}	-7896.0	C_{1111c}	-30520
$C_{33t}\,(C_{22t})$	10.755	$C_{33c}\,(C_{22c})$	10.814
$C_{333t}\,(C_{222t})$	-122.1	$C_{333c}\,(C_{222c})$	113.9
$C_{3333t}\,(C_{2222t})$	-10390	$C_{3333c}\,(C_{2222c})$	-1775.0
$C_{13t}\,(C_{12t})$	4.2728	$C_{13c}\,(C_{12c})$	4.4150
C_{23t}	4.1799	C_{23c}	4.2086
$C_{66t}\,(C_{55t})$	4.7281	$C_{66c}\,(C_{55c})$	4.3592
$C_{6666t}\,(C_{55t})$	-7670.0	$C_{6666c}\,(C_{55c})$	-1498.0

Figures 9, 10 and 11 show load-deflection curves for flat, convex and concave beams with various initial curvatures and fiber waviness ratios. It is observed from the figures that the degree as well as type of initial curvature affects the nonlinear flexural behavior of composite beams. The load-deflection curves of the convex beam are stiffer than those of the concave beam which are less stiff than those of the flat beam for all fiber waviness ratios. This trend is more pronounced as the fiber waviness ratio of the beam is decreased. These results are well expected, since the composite structures with low fiber waviness ratio are more sensitive to fiber reorientation due to deformation than those with high fiber waviness ratio [7]. When flexural loading is applied to a composite beam, some portion of the beam is exposed to tensile stress and the remainder to compressive stress (see figures 19, 20 and 21). The portion of the beam under tensile stress tends to stretch the wavy fibers while the portion under compressive stress tends to increase the waviness of the fibers. This causes fiber reorientation in the composite beam.

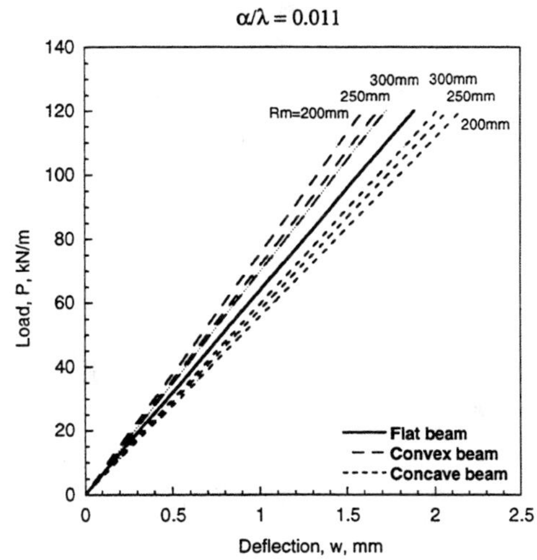

Fig. 9 Predicted load-deflection curves for uniform fiber waviness model with initial curvature and various radii of curvature (a/λ=0.011).

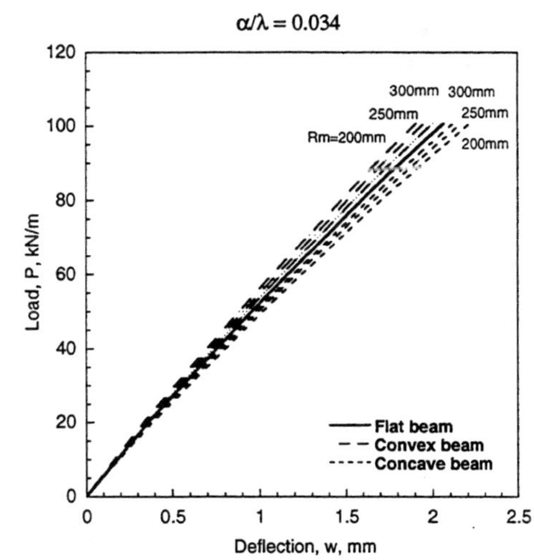

Fig. 10 Predicted load-deflection curves for uniform fiber waviness model with initial curvature and various radii of curvature (a/λ=0.034).

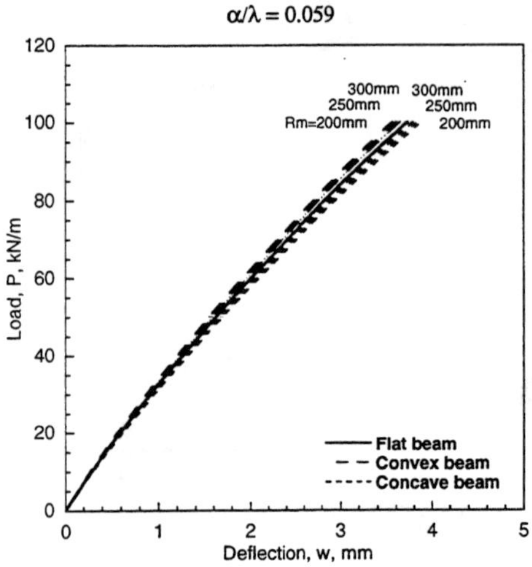

Fig. 11 Predicted load-deflection curves for uniform fiber waviness model with initial curvature and various radii of curvature (a/λ=0.059).

Figure 12 shows the comparison of the load-deflection curves of flat beams for three types of waviness patterns: uniform, graded and localized fiber waviness. It is observed from the figure that the type of fiber waviness significantly affects the nonlinear behavior of thick composites. The load-deflection curve of the uniform fiber waviness model shows the most degradation and nonlinearity among the three models.

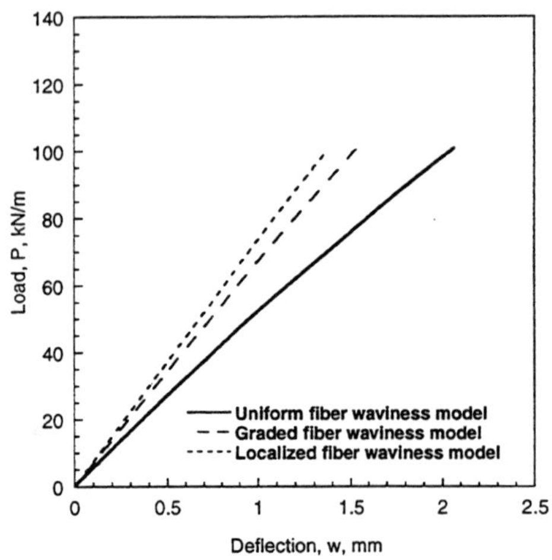

Fig. 12 Predicted load-deflection curves of flat beams for three types of fiber waviness with fiber waviness ratio (a/λ) of 0.034(V_{ct} = 1.0 for graded fiber waviness model, V_{ct} = 1.0 and V_{cl} = 0.5 for localized fiber waviness).

Figures 13 and 14 show predictions of flexural behavior for graded and localized fiber waviness models with an initial curvature. Similar trends of load-deflection curves are observed for graded and localized fiber waviness models as those for uniform fiber waviness model. The localized fiber waviness model shows the least influence of fiber waviness.

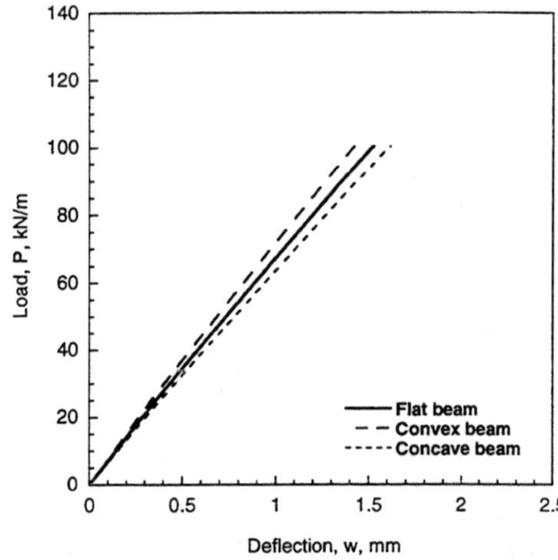

Fig.13 Predicted load-deflection curves for graded fiber waviness model for three types of beams: flat, convex and concave (a/λ = 0.034, V_{ct} = 1.0 and R_m = 200mm).

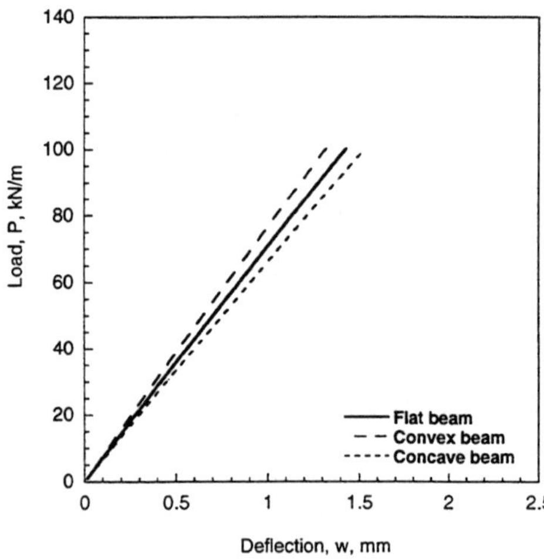

Fig. 14 Predicted load-deflection curves for localized fiber waviness model for three types of beams : flat, convex and concave (a/λ = 0.034, V_{ct} = 1.0, V_{cl} = 0.5 and R_m = 200mm).

Figure 15 shows the comparison among the predicted and experimentally obtained load-deflection curves under flexural loading for flat beams. It is observed from the figure that the predicted deflection for corresponding load by the FEA model is lower than that obtained by the thin slice model [7] and experiment for all fiber waviness ratios. It is also noted that load-deflection curves show more nonlinearity as the fiber waviness ratio increases. The predicted curves by the FEA model show good agreement with predictions by the thin slice model and the experiments.

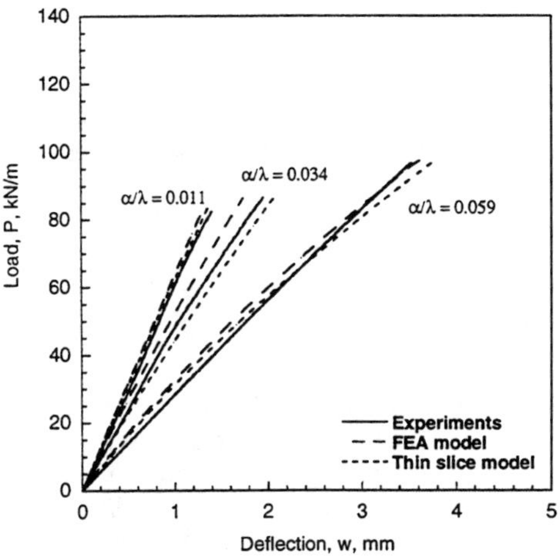

Fig. 15 Predicted and experimentally obtained load-deflection curves for flat uniform fiber waviness model with various fiber waviness ratios.

Figures 16, 17 and 18 show the change in fiber waviness after deformation for three types of beams under the same applied load (P=100kN/m). The x-axis and z-axis in the figures indicate the support span of 140mm and distance from a point on the midplane of one reference end, respectively. It is shown from the figures that the distance between upper peaks of waviness decreases while that between lower peaks increases for both convex and concave beams. It is also observed from the figures that the amplitude of fiber waviness decreases in the end regions whereas it increases in the center region of the beam varying from a maximum at the middle to a minimum at the ends. These trends are more pronounced in the case of convex beams with smaller radius of curvature. The effects of degree of initial curvature on flexural behavior are more significant in the case of concave beams than convex beams.

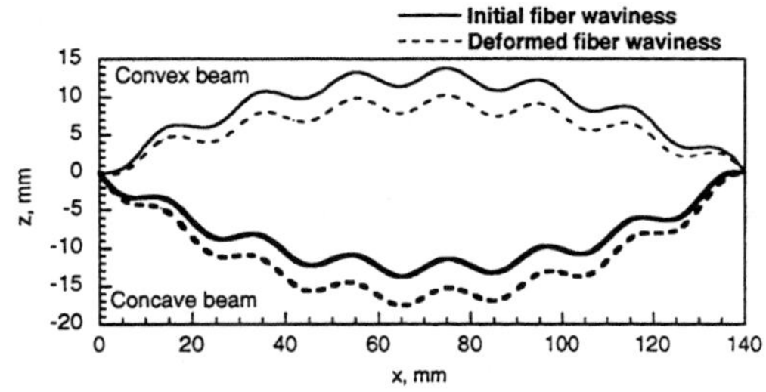

Fig. 16 Initial and deformed fiber waviness after deformation (P=100kN/m, a/λ = 0.059, R_m = 200mm).

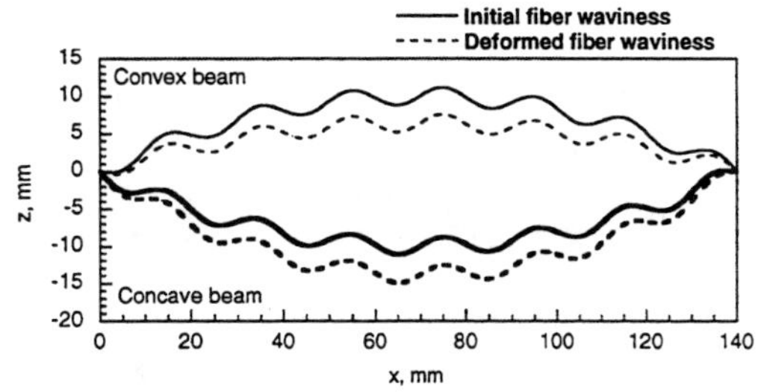

Fig. 17 Initial and deformed fiber waviness after deformation (P=100kN/m, a/λ = 0.059, R_m = 250mm).

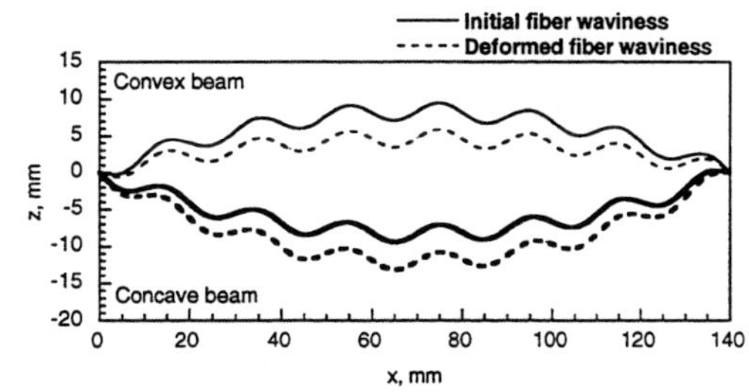

Fig. 18 Initial and deformed fiber waviness after deformation (P=100kN/m, a/λ = 0.059, R_m = 300mm).

Figure 19 shows results of stress analysis for the flat uniform fiber waviness model obtained by FEA for P=50kN/m. A schematic drawing showing the initial fiber waviness and contour plots for four stress components (σ_x, σ_z, σ_{xz} and σ_{eff}) are given. It is noted from Fig. 19-(b) that the neutral plane has shifted below the midplane due to geometric nonlinearity associated with change of fiber waviness during deformation. It is also observed from the figure that the axial stress is the dominant stress component for flexural loading. The maximum and minimum axial stresses are generated at the top and bottom sides of the flat beam. Figure 19-(d) shows the existence of shear stress due to shear coupling effect. However, the magnitudes of shear stress are small.

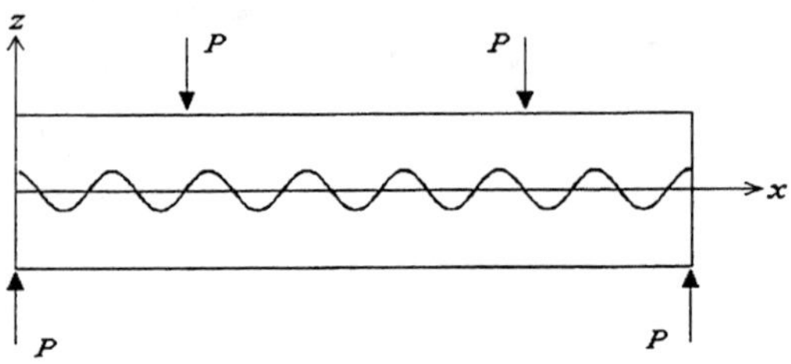

(a)

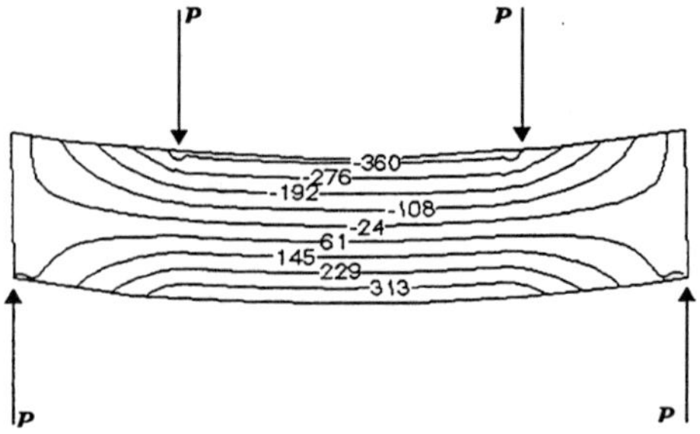

(b)

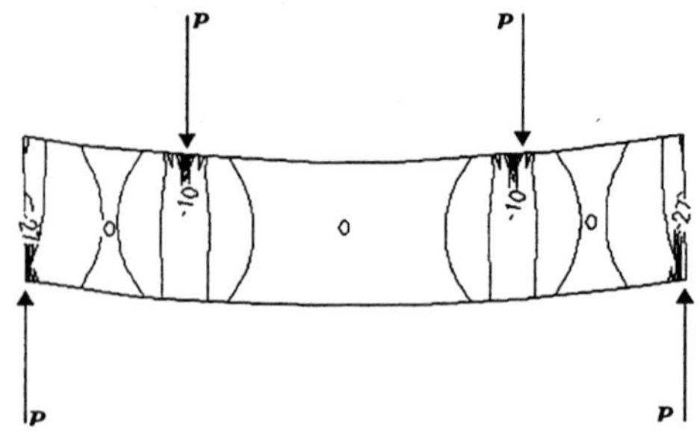

(c)

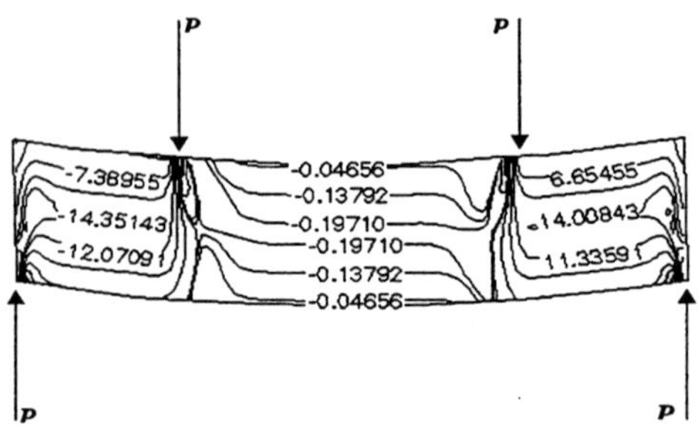

(d)

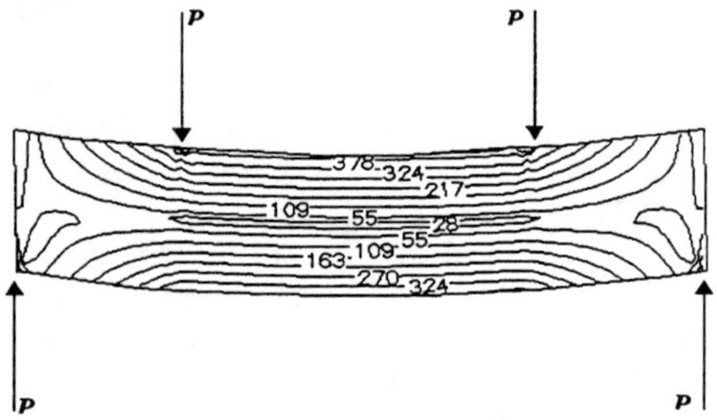

(e)

Fig. 19 Stress analyses of flat beam with uniform fiber waviness (applied load, P = 50kN/m) : (a) schematic drawing of initial fiber waviness ratio (a/λ) of 0.034, (b) contour plot of axial stress (σ_x, MPa), (c) contour plot of lateral stress (σ_z, MPa), (d) contour plot of shear stress (σ_{xz}, MPa), (e) contour plot of effective stress (σ_{eff}, MPa).

Similar plots for uniform fiber waviness models with an initial curvature are shown in figures 20 and 21. It is very interesting to observe that the maximum and maximum hoop stresses are different between concave and convex beams for the same applied load. The higher values are obtained for convex beams with the same initial curvature, which indicates that the sign of initial curvature affects the stress distribution in the beams.

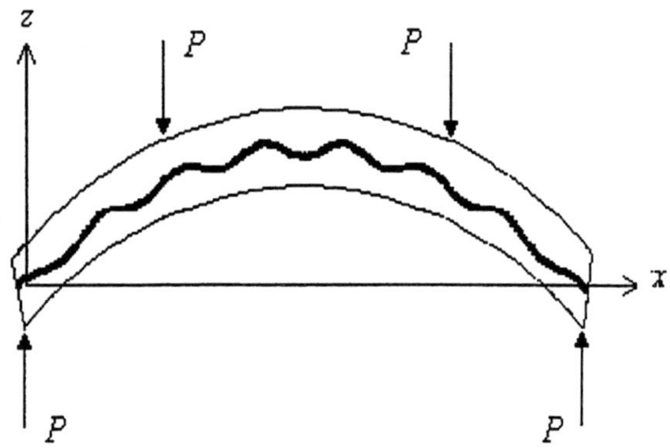

(a)

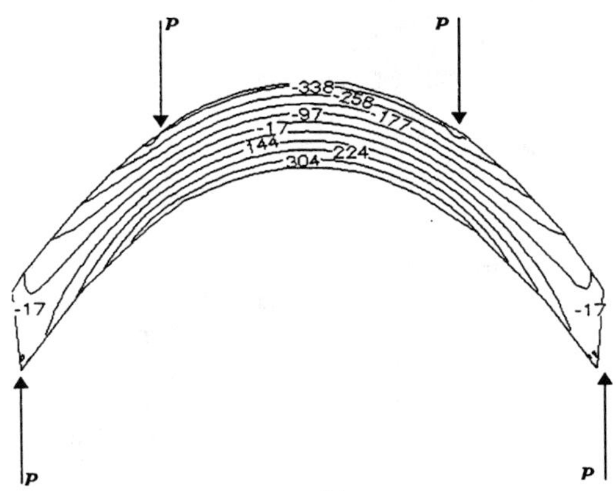

(b)

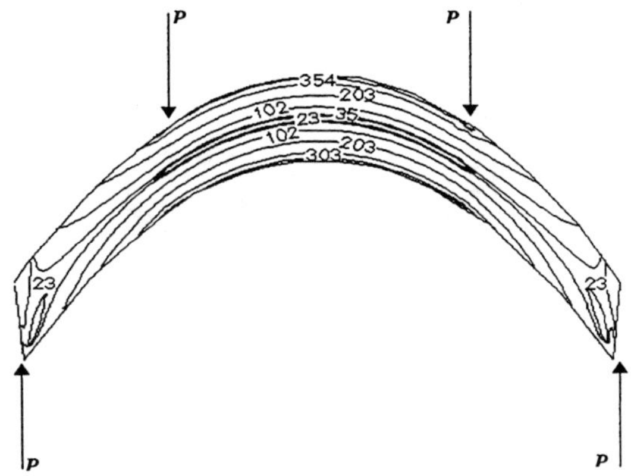

(e)

Fig. 20 Stress analyses of convex beam with uniform fiber waviness model (applied load, P = 50kN/m) : (a) schematic drawing of initial fiber waviness ratio (a/λ) of 0.034 and radius of curvature (R_m) of 200mm, (b) contour plot of hoop stress (σ_ϕ, MPa), (c) contour plot of radial stress (σ_r, MPa), (d) contour plot of shear stress ($\sigma_{r\phi}$, MPa), (e) effective stress (σ_{eff}, MPa).

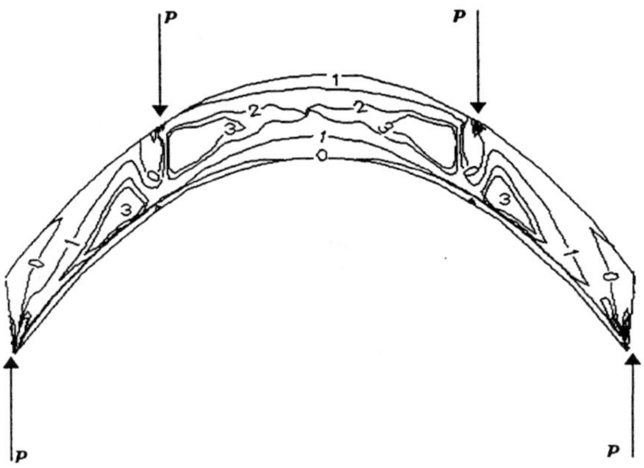

(c)

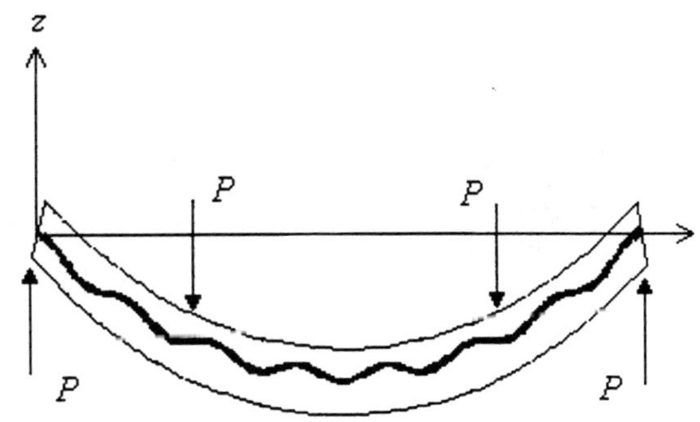

(a)

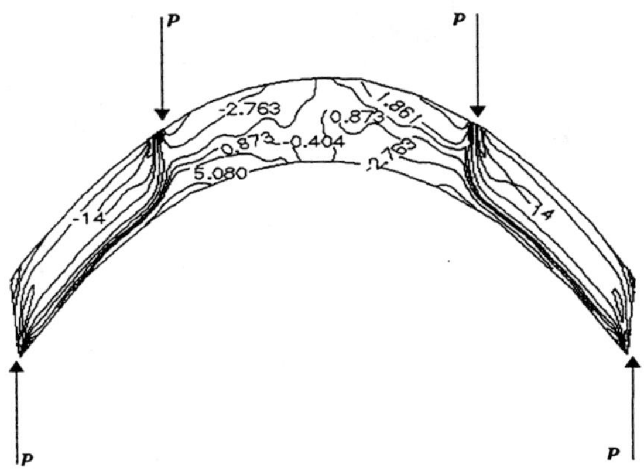

(d)

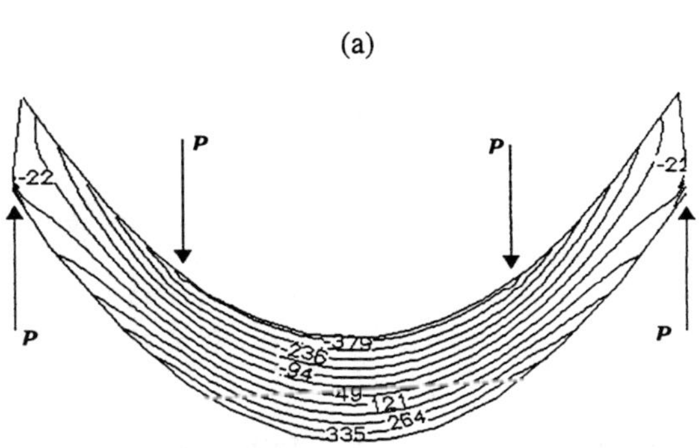

(b)

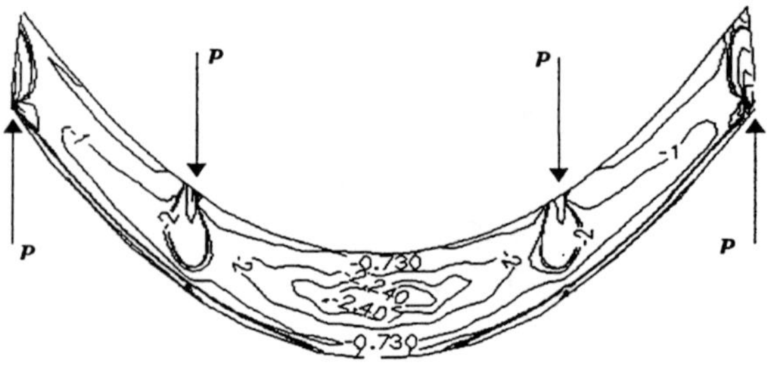

(c)

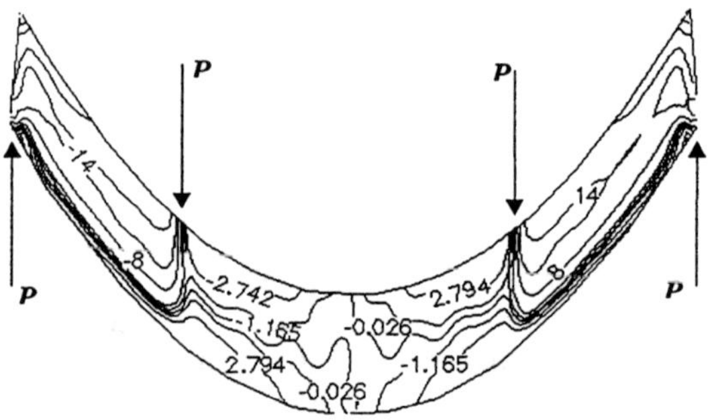

(d)

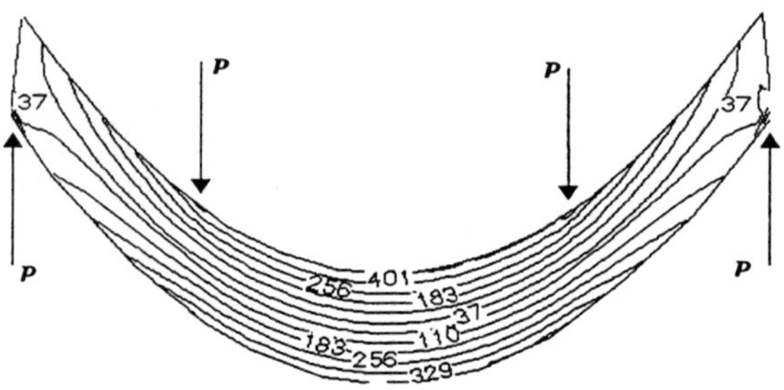

(e)

Fig. 21 Stress analyses of concave beam with uniform fiber waviness model (applied load, P = 50kN/m) : (a) schematic drawing of initial fiber waviness ratio (a/λ) of 0.034 and radius of curvature (R_m)of 200mm, (b) contour plot of hoop stress (σ_ϕ, MPa), (c) contour plot of radial stress (σ_r, MPa), (d) contour plot of shear stress ($\sigma_{r\phi}$, MPa), (e) effective stress (σ_{eff}, MPa).

SUMMARY AND CONCLUSIONS

A FEA model was proposed to study the nonlinear elastic flexural behaviors of thick composites with three types of fiber waviness (uniform, graded and localized) and initial curvature (flat, convex and concave beam).

Energy density, iterative mapping and incremental analysis were used to incorporate the material and geometric nonlinearities due to fiber waviness.

It was found that the degree of fiber waviness and initial curvature as well as the type of fiber waviness in composite materials significantly affected the nonlinear flexural behavior of composites and the shape of fiber waviness during the deformation. It was also found that the degree of initial curvature affected more significantly on the flexural behavior of composites beams in the case of convex than concave beams.

The predictions of flexural behavior of flat thick composite beams with uniform fiber waviness using a FEA model were compared with analytical and experimental results. The predictions based on the FEA model were in good agreement with the analytical and experimental results.

In addition, stress analyses were conducted on the uniform fiber waviness model with different types of curvatures. It was concluded that the sign of initial curvature affected the stress distribution. Axial and hoop stresses were the dominant stress components in flat, concave, and convex beams, respectively, under four point flexural loading.

REFERENCES

[1] Shuart, M. J., *AIAA Journal*, Vol. 27, No. 9, 1989, pp. 1274-1279.
[2] Telegadas, H. K. and Hyer, M. W., *Journal of Reinforced Plastics and Composites*, Vol. 11, No. 2, 1992, pp. 127-145.
[3] Chou, T.-W. and Takahashi, K., *Composites*, Vol. 18, No. 1, 1987, pp. 25-33.
[4] Hsiao, H. M. and Daniel, I. M., *Journal of Engineering Materials and Technology*, Vol. 118, 1996, pp. 561-570.
[5] Hsiao, H. M. and Daniel, I. M., *Composites Part A*, Vol. 27A, 1996, pp. 931-941.
[6] Hsiao, H. M. and Daniel, I. M., *Journal of Advanced Materials*, Vol. 26, 1996, pp. 19-26.
[7] Chun, H.-J., Daniel, I. M. and Shin, J.-Y., *A Collection of Technical Papers, AIAA/ASME/ASCE/AHS/ASC Structures, Structural Dynamics, and Materials Conference and Exhibit, St. Louis, MO, April 12-15*, 1999, Vol. 1, pp. 467-474.
[8] Kedward, K. T., Wilson, R. S. and McLean, S. K., *Composites*, Vol. 20, No.6, 1989, pp. 527-536.
[9] Ganapathi, M., Polit, O. and Touratier, M., *International Journal for Numerical Methods in Engineering*, Vol. 39, 1996, pp. 3453-3474.

MIXED MODE INTERFACE CRACKS IN A BI-MATERIAL HALF PLANE AND A BI-MATERIAL STRIP

Haiying Huang[1], **George A. Kadomateas**[2], and **Valeria La Saponara**[3]

School of Aerospace Engineering, Georgia Institute of Technology
Atlanta, Georgia 30332-0150, USA

ABSTRACT

This paper presents a method for determining the dislocation solution in a bi-material half plane and a bi-material infinite strip, which is subsequently used to obtain the mixed-mode stress intensity factors for a corresponding bi-material interface crack. First, the dislocation solution in a bi-material infinite plane is summarized. An array of surface dislocations is then distributed along the free boundary of the half plane and the infinite strip. The dislocation densities of the aforementioned surface dislocations are determined by satisfying the traction-free boundary conditions. After the dislocation solution in the finite domain is achieved, the mixed-mode stress intensity factors for interface cracks are calculated based on the continuous dislocation technique. Results are compared with analytical solution for homogeneous anisotropic media.

INTRODUCTION

The study of interfacial phenomena in bimaterials is currently a topic of considerable interest to the applied mechanics community because the overall performance of many engineered-material systems, such as composite structures, electronic packaging, and thin film constructions, are largely depended on the toughness of the interface. Accurate stress intensity factor calculations are essential in the prediction of interface crack growth rates and fracture strength. A very effective method of determining stress intensities is the continuous dislocation technique. The cornerstone of this method is the fundamental solution of a dislocation in the corresponding configuration. Eshelby et.al. (1953) and Stroh(1958) are among the pioneers who presented an analytical solution for a dislocation in general anisotropic materials. Following their work, Ting (1986), and Qu and Li (1991) studied the classical problem of a dislocation situated at the interface between two anisotropic elastic half planes and obtained a solution to the dislocation problem. Atkinson and Eftaxiopoulos (1991) also achieved the solution for a dislocation in an anisotropic half plane and bimaterial, using the basic formulation of Stroh and appropriate image systems.

All these analytic dislocation solutions are restricted to bimaterial infinite planes. The solution for a bimaterial media with traction-free boundaries such as the half-plane and the infinite strip are not available in literature so far, although some numerical methods for homogeneous materials were presented in the past. Civelek and Erdogan (1982) developed a numerical method to calculate the dislocation solution in an isotropic infinite strip by superposing the infinite plane with an additional elastic field, which is expressed by an Airy stress function with Fourier transformation. Suo (1990) and Suo and Hutchinson (1990) extended this method to orthotropic materials and calculated the mixed-mode stress intensity factors for an infinite strip with semi infinite cracks subjected to edge bending. A surface dislocation model was proposed by Jagannagham and Marcinkowski(1979) to calculate the stress fields of a finite body subjected to either

---

[1] Ph.D., FMC Corporation, APSD, Orlando, FL 32809

[2] Professor

[3] Ph.D. candidate

applied stresses or an internal stress. Huang and Kardomateas (1999) developed a method to calculate the stress fields of a dislocation in homogeneous anisotropic infinite strips and applied the solution to calculate the stress intensity factors for both central cracks and edge cracks in an infinite strip.

In this paper, the analytic solution for a dislocation in an anisotropic bimaterial infinite plane is summarized first. The stress fields of a dislocation in the bi-material half plane and the bi-material strip are subsequently obtained by distributing arrays of dislocations along the traction free boundaries of these two finite domains. These dislocation solutions are then applied to calculate the stress intensity factors of an interface crack in both a bi-material half plane and a bi-material strip. Some results of practical interest are presented.

FORMULATION

Dislocation Solution in a Bi-material Infinite Plane:

The analytical solution for dislocations in a bi-material infinite plane has several different versions. Almost all of them originate from Stroh's formulation. Combining the solutions presented by Ting(1986), Qu and Li(1991), and Atkinson and Eftaxiopoulos(1991), we present first a concise summary of this elegant analytical solution for a dislocation in anisotropic infinite plane.

The stress potentials for an anisotropic media can be expressed as:

$$\{\phi_i\} = \mathbf{B}\{f_\alpha(z_\alpha)\} + \bar{\mathbf{B}}\{\bar{f}_\alpha(\bar{z}_\alpha)\}, \quad i,\alpha = 1,\ 2,\ 3. \tag{1}$$

The 3×3 matrix $\mathbf{B}$ is introduced by Ting as:

$$\mathbf{B} = (\mathbf{R}^T + \mathbf{pT})\mathbf{A}\ . \tag{2}$$

$\mathbf{p} = \{p_1, p_2, p_3\}^T$ and $\mathbf{A} = [\mathbf{a}_1, \mathbf{a}_2, \mathbf{a}_3]$ are the eigenvector and eigenvalue matrix of the following equation, respectively:

$$\{\mathbf{Q} + \mathbf{p}(\mathbf{R} + \mathbf{R^T}) + \mathbf{p^2T}\}\mathbf{a} = \mathbf{0}\ , \tag{3}$$

where

$$\mathbf{Q} = [Q_{ik}] = [C_{i1k1}]\,,\quad \mathbf{R} = [R_{ik}] = [C_{i1k2}]\,, \mathbf{T} = [T_{ik}] = [C_{i2k2}]\,, \qquad i,k = 1,2,3. \tag{4}$$

For plane strain and plane stress problem, one of the eigenvetor will be zero and $\mathbf{A}$ is reduced to a 2×2 matrix. Denoting the eigenvectors with positive imaginary part as p_α, $\alpha = 1,\ 2$, z_α in equation (1) are defined as following:

$$z_\alpha = x_1 + p_\alpha x_2 = \frac{1 - ip_\alpha}{2}z + \frac{1 + ip_\alpha}{2}\bar{z}, \qquad \alpha = 1,2 \tag{5}$$

The stress components can be calculated from equation (1) as:

$$\sigma_{i1} = -\frac{\partial \phi_i}{\partial x_2} = -\mathbf{B} \ll p_\alpha \gg \{f'_\alpha(z_\alpha)\} - \bar{\mathbf{B}} \ll \bar{p}_\alpha \gg \{\bar{f}'_\alpha(\bar{z}_\alpha)\} \tag{6a}$$

$$\sigma_{i2} = \frac{\partial \phi_i}{\partial x_1} = \mathbf{B}\{f'_\alpha(z_\alpha)\} + \bar{\mathbf{B}}\{\bar{f}'_\alpha(\bar{z}_\alpha)\}, \quad i,\ \alpha = 1,\ 2. \tag{6b}$$

The double angle brackets stand for diagonal matrix, i.e.

$$\ll p_\alpha \gg = \begin{bmatrix} p_1 & 0 \\ 0 & p_2 \end{bmatrix}\ . \tag{7}$$

Assuming the dislocation $\mathbf{b} = \{b_1, b_2\}^T$ is located at $z_0 = x_{10} + ix_{20}$ in the upper half plane (media (1)) of the bi-material infinite plane(as shown in Figure 1), the functions $f_\alpha(z_\alpha)$ in equation (1) and (6) can be found in Atkinson and Eftaxiopoulos(1991):

$$f_\alpha^{(1)}(z_\alpha^{(1)}) = \frac{1}{4\pi}\mathbf{M}^{(1)}\mathbf{D}^{(1)}\mathbf{b}^{(1)}\ln(z_\alpha^{(1)} - z_{0\alpha}^{(1)}) + \frac{1}{4\pi}\mathbf{E}\ln(z_\alpha^{(1)} - \bar{\mathbf{z}}_0^{(1)})\ , \tag{8}$$

$$f_\alpha^{(2)}(z_\alpha^{(2)}) = \frac{1}{4\pi}\mathbf{G}\ln(z_\alpha^{(2)} - \mathbf{z}_0^{(1)})\ , \tag{9}$$

where $\mathbf{z_0}^{(1)} = \left\{z_{01}^{(1)}, z_{02}^{(1)}\right\}^T$, $\mathbf{M}^{(1)} = \mathbf{B}^{(1)^{-1}}$, and $\mathbf{D}^{(1)} = -2i(\mathbf{A}^{(1)}\mathbf{M}^{(1)} - \bar{\mathbf{A}}^{(1)}\bar{\mathbf{M}}^{(1)})^{-1}$. The superscript (1) and (2) indicate material (1) and (2). $\mathbf{E}$ and $\mathbf{G}$ are constants, depending on the dislocation density $\mathbf{b}$ and the elastic properties of media (1) and (2). They can determined by solving the following linear equations:

$$\begin{bmatrix} -\bar{\mathbf{A}}^{(1)} & \mathbf{A}^{(2)} \\ -\bar{\mathbf{B}}^{(1)} & \mathbf{B}^{(2)} \end{bmatrix} \begin{Bmatrix} \bar{\mathbf{E}} \\ \mathbf{G} \end{Bmatrix} = \begin{Bmatrix} \bar{\mathbf{A}}^{(1)} \ll \mathbf{M}^{(1)}\mathbf{D}^{(1)}\mathbf{b} \gg \\ \bar{\mathbf{B}}^{(1)} \ll \mathbf{M}^{(1)}\mathbf{D}^{(1)}\mathbf{b} \gg \end{Bmatrix} . \tag{10}$$

Detailed derivation of equation (10) can be found in Atkinson and Eftaxiopoulos(1991). Although in that paper it was not specifically stated that their solution is applicable to the interface dislocation, our calculation yielded the same results from both Atkinson and Eftaxiopoulos's (1991) and Qu and Li's (1991) method for interface dislocations.

Because of the linearity of the stress fields of dislocations in the infinite bimaterial plane, we can express the stress components at $z = x_1 + ix_2$ due to a dislocation $\mathbf{b} = \{b_1, b_2\}^T$ at $z_0 = x_{10} + ix_{20}$ as:

$$\sigma_{ij}(x_1, x_2) = \mathbf{F}(x_1, x_2, x_{10}, x_{20})\mathbf{b}(x_{10}, x_{20}) , \tag{11}$$

where

$$\mathbf{F}(x_1, x_2, x_{10}, x_{20}) = [f_{1ij}(x_1, x_2, x_{10}, x_{20}), f_{2ij}(x_1, x_2, x_{10}, x_{20})] ,$$

and

$$\mathbf{b}(x_{10}, x_{20}) = \{b_1(x_{10}, x_{20}), b_2(x_{10}, x_{20})\}^T .$$

The physical meaning of f_{1ij} and f_{2ij} are stress component σ_{ij} due to unit dislocations $b_1 = 1$ and $b_2 = 1$, respectively. Therefore, we can calculate f_{1ij} and f_{2ij} from equation (6) by setting $\mathbf{b} = \{1, 0\}^T$ and $\mathbf{b} = \{0, 1\}^T$.

Dislocation Solution in a Bi-Material Half Plane and a Bi-Material Strip:

Next, let's consider the numerical solution for a dislocation in a bimaterial half plane and a bimaterial strip. The basic idea of calculating the dislocation solution in the half plane and the strip is to apply arrays of dislocations along the free boundaries of the half plane and the infinite strip. The densities of the dislocation arrays are determined in such a way that the traction forces along the boundaries due to the single dislocation and the dislocation arrays are zero, as shown in Figure 2. We will derive the dislocation solution in a bimaterial infinite strip first, it will be clear later that the solution can be easily reduced to the half plane solution.

As shown in Figure 2, the bimaterial infinite strip consists of two anisotropic half planes, with material (1) lay on top of material (2). The thickness of material (1) is H and the thickness of material (2) is h. The origin of the coordinate system is located at the interface. A dislocation $\mathbf{b} = \{b_1, b_2\}^T$ is located at an arbitrary point $z_0 = x_{10} + ix_{20}$. The geometry of a dislocation in the infinite strip can be decomposed into two geometries. The first one is a single dislocation located in the bimaterial infinite plane. The two dashed lines stand for the boundaries of the infinite strip, which are supposed to be traction free. The traction forces along the dashed lines due to the single dislocation can be calculate from equation(11):

$$\sigma_{ij}^{(s)}(x_1^{(1)}, H) = \mathbf{F}(x_1^{(1)}, H, x_{10}, x_{20})\mathbf{b}(x_{10}, x_{20}) , \tag{12a}$$

$$\sigma_{ij}^{(s)}(x_1^{(2)}, -h) = \mathbf{F}(x_1^{(2)}, -h, x_{10}, x_{20})\mathbf{b}(x_{10}, x_{20}) , \tag{12b}$$

where $ij = 12, 22$.

The second geometry is the infinite plane with two arrays of dislocations along the supposed-to-be boundaries of the infinite strip. To satisfy the traction free boundary conditions of the infinite strip, the traction forces along the dashed lines in the second geometry should be the opposite of the traction forces in the first geometry. Suppose that the dislocation arrays are distributed from $-\infty$ to ∞, then the tractions along the dashed lines due to the two dislocation arrays can be calculated as:

$$\sigma_{ij}(x_1^{(1)}, H) = \int_{-\infty}^{\infty} \mathbf{F}(x_1^{(1)}, H, t^{(1)}, H)\mathbf{b}(t^{(1)}, H)dt^{(1)} +$$

$$\int_{-\infty}^{\infty} \mathbf{F}(x_1^{(1)}, H, t^{(2)}, -h)\mathbf{b}(t^{(2)}, -h)dt^{(2)} = -\sigma_{ij}^{(s)}(x_1^{(1)}, H) , \tag{13a}$$

and

$$\sigma_{ij}(x_1^{(2)}, -h) = \int_{-\infty}^{\infty} \mathbf{F}(x_1^{(2)}, -h, t^{(1)}, H)\mathbf{b}(t^{(1)}, H)dt^{(1)} +$$

$$\int_{-\infty}^{\infty} \mathbf{F}(x_1^{(2)}, -h, t^{(2)}, -h)\mathbf{b}(t^{(2)}, -h)dt^{(2)} = -\sigma_{ij}^{(s)}(x_1^{(2)}, -h) \ , \tag{13b}$$

where $ij = 12,\ 22$. $\sigma_{ij}^{(s)}(x_1^{(1)}, H)$ and $\sigma_{ij}^{(s)}(x_1^{(2)}, -h)$ are defined in equation(12).

Equations (13) is a set of singular integral equations. Gaussian quadrature is adopted to solve the integrations numerically. Since the single dislocation is in self-equilibrium, the traction forces $\sigma_{ij}^{(s)}(x_1^{(1)}, H)$ and $\sigma_{ij}^{(s)}(x_1^{(2)}, -h)$ vanish when $t^{(1)}, t^{(2)} \to -\infty, +\infty$. As a result, the dislocation densities along the boundaries go to zero, too. Therefore, we can ignore the dislocations located at $x_1 > d_1$ and $x_1 < -d_1$ for boundary $x_1 = H$ and $x_1 > d_2$ and $x_1 < -d_2$ for boundary $x_1 = -h$, where d_1 and d_2 are values large enough compared to H and $-h$. Usually $d_1 = 50H$ and $d_2 = 50h$ are accurate enough. Thus, equation (13) becomes:

$$\sigma_{ij}(x_1^{(1)}, H) = \int_{-d_1}^{d_1} \mathbf{F}(x_1^{(1)}, H, t^{(1)}, H)\mathbf{b}(t^{(1)}, H)dt^{(1)} +$$

$$\int_{-d_1}^{d_1} \mathbf{F}(x_1^{(1)}, H, t^{(2)}, -h)\mathbf{b}(t^{(2)}, -h)dt^{(2)} = -\sigma_{ij}^{(s)}(x_1^{(1)}, H) \ , \tag{14a}$$

and

$$\sigma_{ij}(x_1^{(2)}, -h) = \int_{-d_1}^{d_1} \mathbf{F}(x_1^{(2)}, -h, t^{(1)}, H)\mathbf{b}(t^{(1)}, H)dt^{(1)} +$$

$$\int_{-d_2}^{d_2} \mathbf{F}(x_1^{(2)}, -h, t^{(2)}, -h)\mathbf{b}(t^{(2)}, -h)dt^{(2)} = -\sigma_{ij}^{(s)}(x_1^{(2)}, -h) \ . \tag{14b}$$

We can normalize equations (14) as follows:

$$\tilde{t}^{(1)} = \frac{t^{(1)}}{d_1} \quad \text{and} \quad \tilde{t}^{(2)} = \frac{t^{(2)}}{d_2}$$

And now rewrite equations (14) in a form that admits Gaussian quadrature,

$$\sigma_{ij}(x_1^{(1)}, H) = \pi d_1 [\frac{1}{\pi} \int_{-1}^{1} \mathbf{F}_{ij}(x_1^{(1)}, H, \tilde{t}^{(1)}, H)\mathbf{b}(\tilde{t}^{(1)}, H)d\tilde{t}^{(1)}] +$$

$$\pi d_2 [\frac{1}{\pi} \int_{-1}^{1} \mathbf{F}_{ij}(x_1^{(1)}, H, \tilde{t}^{(2)}, -h)\mathbf{b}(\tilde{t}^{(2)}, -h)d\tilde{t}^{(2)}] = -\sigma_{ij}^{(s)}(x_1^{(1)}, H) \ , \tag{15a}$$

and

$$\sigma_{ij}(x_1^{(2)}, -h) = \pi d_1 [\frac{1}{\pi} \int_{-1}^{1} \mathbf{F}_{ij}(x_1^{(2)}, -h, t^{(1)}, H)\mathbf{b}(\tilde{t}^{(1)}, H)d\tilde{t}^{(1)}] +$$

$$\pi d_2 [\frac{1}{\pi} \int_{-1}^{1} \mathbf{F}_{ij}(x_1^{(2)}, -h, \tilde{t}^{(2)}, -h)\mathbf{b}(t^{(2)}, -h)d\tilde{t}^{(2)}] = -\sigma_{ij}^{(s)}(x_1^{(2)}, -h) \ . \tag{15b}$$

It is reasonable to assure that the dislocation densities of the dislocation arrays are zero at $x_1 = -d_1,\ d_1, -d_2, d_2$ as long as $d_1,\ d_2$ are large enough. Accordingly, the dislocation densities may be written as:

$$b(\tilde{t}^{(1)}, H) = W^{(1)}(\tilde{t}^{(1)})\tilde{b}(\tilde{t}^{(1)}, H); \quad W^{(1)}(\tilde{t}^{(1)}) = \sqrt{1 - \tilde{t}^{(1)2}} \ , \tag{16a}$$

and

$$b(\tilde{t}^{(2)}, -h) = W^{(2)}(\tilde{t}^{(2)})\tilde{b}(\tilde{t}^{(2)}, -h); \quad W^{(2)}(\tilde{t}^{(2)}) = \sqrt{1 - \tilde{t}^{(2)2}} \ . \tag{16b}$$

Substituting equations (16) into equations (15), the numerical form of the singular integral equations can be reduced to a set of algebraic equations having the following matrix forms:

$$[\tilde{F}_{kl}] \ll [W_l^{(1)}, W_l^{(1)}, W_l^{(2)}, W_l^{(2)}] \gg \tilde{\mathbf{B}} = \begin{Bmatrix} -\sigma_{22}^{(s)}(x_{1,k}^{(1)}, -h) \\ -\sigma_{12}^{(s)}(x_{1,k}^{(1)}, -h) \\ -\sigma_{22}^{(s)}(x_{1,k}^{(2)}, H) \\ -\sigma_{12}^{(s)}(x_{1,k}^{(2)}, H) \end{Bmatrix} \tag{17}$$

where

$$[\tilde{F}_{kl}] = \begin{bmatrix} \pi d_1 \mathbf{F}_{22}(x_{1,k}^{(1)}, H, t_l^{(1)}, H) & \pi d_2 \mathbf{F}_{22}(x_{1,k}^{(1)}, H, t_l^{(2)}, -h) \\ \pi d_1 \mathbf{F}_{12}(x_{1,k}^{(1)}, H, t_l^{(1)}, H) & \pi d_2 \mathbf{F}_{12}(x_{1,k}^{(1)}, H, t_l^{(2)}, -h) \\ \pi d_1 \mathbf{F}_{22}(x_{1,k}^{(2)}, -h, t_l^{(1)}, H) & \pi d_2 \mathbf{F}_{22}(x_{1,k}^{(2)}, -h, t_l^{(2)}, -h) \\ \pi d_1 \mathbf{F}_{12}(x_{1,k}^{(2)}, -h, t_l^{(1)}, H) & \pi d_2 \mathbf{F}_{12}(x_{1,k}^{(2)}, -h, t_l^{(2)}, -h) \end{bmatrix}, \tag{18}$$

and

$$\tilde{\mathbf{B}} = \begin{Bmatrix} \tilde{\mathbf{b}}(\tilde{t}_l^{(1)}, H) \\ \tilde{\mathbf{b}}(\tilde{t}_l^{(2)}, -h) \end{Bmatrix}. \tag{19}$$

the integration and collocation points are calculated based on Gaussian quadrature,

$$\tilde{t}_l^{(1)} = \cos(\frac{\pi l}{N_1+1}); \quad \tilde{x}_{1,k}^{(1)} = \cos[\frac{\pi(2k-1)}{2(N_1+1)}]; \quad W_l^{(1)} = \frac{1-\tilde{t}_l^{(1)2}}{N_1+1}$$
$$l, k = 1, 2, \ldots, N_1$$

and

$$\tilde{t}_l^{(2)} = \cos(\frac{\pi l}{N_2+1}); \quad \tilde{x}_{1,k}^{(2)} = \cos[\frac{\pi(2k-1)}{2(N_2+1)}]; \quad W_l^{(2)} = \frac{1-\tilde{t}_l^{(2)2}}{N_2+1}$$
$$l, k = 1, 2, \ldots, N_2$$

where N_1, N_2 are the numbers of integration and collocation points along the two traction-free boundaries of the bi-material infinite strip. Consequently,

$$x_{1,k}^{(1)} = d_1 \tilde{x}_{1,k}^{(1)}; \quad t_i^{(1)} = d_1 \tilde{t}_l^{(1)}; \quad x_{1,k}^{(2)} = d_2 \tilde{x}_{1,k}^{(2)}; \text{ and } t_i^{(2)} = d_2 \tilde{t}_l^{(2)}. \tag{20}$$

¿From equation (17), we can calculate the dislocation densities $\tilde{\mathbf{B}}$, which are related to the single dislocation $\mathbf{b}$. Denote the dislocation densities along the boundaries of the infinite strip as $\tilde{\mathbf{B}}_1$ due to $\mathbf{b} = \{1, 0\}^T$, and as $\tilde{\mathbf{B}}_2$ due to $\mathbf{b} = \{0, 1\}^T$. Superposing the two elastic fields in Figure 2, we obtained the stress fields for a dislocation $\mathbf{b}_0 = \{b_1, b_2\}^T$ located at $z_0 = x_{10} + ix_{20}$ in the infinite strip:

$$\sigma_{ij}(x_1, x_2) = b_1(x_{10}, x_{20}) \tilde{f}_{1ij}(x_1, x_2, x_{10}, x_{20}) + b_2(x_{10}, x_{20}) \tilde{f}_{2ij}(x_1, x_2, x_{10}, x_{20}), \tag{21}$$

where:

$$\tilde{f}_{1ij}(x_1, x_2, x_{10}, x_{20},) = f_{1ij}(x_1, x_2, x_{10}, x_{20}) + [\tilde{F}_{ij}(x_1, x_2)] \ll [W_l^{(1)}, W_l^{(1)}, W_l^{(2)}, W_l^{(2)}] \gg \tilde{\mathbf{B}}_1, \tag{22a}$$

$$\tilde{f}_{2ij}(x_1, x_2, x_{10}, x_{20},) = f_{2ij}(x_1, x_2, x_{10}, x_{20}) + [\tilde{F}_{ij}(x_1, x_2)] \ll [W_l^{(1)}, W_l^{(1)}, W_l^{(2)}, W_l^{(2)}] \gg \tilde{\mathbf{B}}_2, \tag{22b}$$

and

$$[\tilde{F}_{ij}(x_1, x_2)] = [\pi d_1 \mathbf{F}_{ij}(x_1, x_2, t_i^{(1)}, H), \pi d_2 \mathbf{F}_{ij}(x_1, x_2, t_i^{(2)}, -h)]. \tag{23}$$

Only one dislocation array is needed for the half plane because there is only one traction-free boundary. The stress fields for a dislocation in the bimaterial strip (eq. (21)) can be reduced to those for the half plane by setting either N_1 or N_2 equal to zero.

Stress Intensity Factors for an Interface Bi-Material Crack in the Half Plane and the Strip:

In this section, the dislocation solution developed above will be applied to calculate the stress intensity factors for interface cracks in a bi-material half plane and a bi-material infinite strip. Only interface cracks are studied in this paper because they are of extensive practical importance. However, the method described below can be very easily modified for other types of cracks, such as a slant crack with an arbitrary angle to the interface.

A bimaterial infinite strip with an interface crack of length $2a$ is shown in Figure 3. The origin of the coordinate system is located at the middle point of the crack. The external loads T_{12} & T_{22} are distributed uniformly along the crack surfaces. The crack can be simulated as a series of interface dislocations with dislocation densities $\mathbf{b}(t, 0)$. The tractions along the crack surfaces due to the dislocation series are:

$$\sigma_{ij}(x, 0)^{(d)} = \int_{-a}^{a} [b_1(t, 0) \tilde{f}_{1ij}(x_1, 0, t, 0) + b_2(t, 0) \tilde{f}_{2ij}(x_1, 0, t, 0)] dt, \quad ij = 12, 22 \tag{24}$$

which should be the equivalent to the external loads T_{12} and T_{22}. Normalizing the coordinates by $\tilde{x}_1 = \frac{x_1}{a}$, $\tilde{t} = \frac{t}{a}$, and using Gaussian quadrature, we have $2(N-1)$ linear algebraic equations:

$$\pi a \left[\sum_{l=1}^{N} \tilde{W}_l \tilde{b}_1(t_l,0) \tilde{f}_{1ij}(x_{1k},0,t_l,0) + \sum_{l=1}^{N} \tilde{W}_l \tilde{b}_2(t_l,0) \tilde{f}_{2ij}(x_{1k},0,t_l,0) \right] = T_{ij}(x_{1k},0),$$

$$ij = 12, 22, \quad k = 1, 2, \ldots, N-1. \tag{25}$$

where $\tilde{t}_l = \cos(\frac{\pi(2l-1)}{2N})$; $\tilde{x}_{1k} = \cos(\frac{\pi k}{N})$; and $W_l = \frac{1}{N}$. Also, $t_l = a\tilde{t}_l$, $x_{1k} = a\tilde{x}_{1k}$, and $\tilde{b}_i(\tilde{t},0) = \frac{b_i(\tilde{t},0)}{\sqrt{1-\tilde{t}^2}}$, $i = 1,2$. In addition, the requirement that the crack surfaces physically come together at both ends imposes two additional equations:

$$\sum_{l=1}^{N} W_l \tilde{b}_1(\tilde{t}_l) = 0; \quad \sum_{l=1}^{N} W_l \tilde{b}_2(\tilde{t}_l) = 0; \tag{26}$$

Equation (25) and (26) enable us to calculate the dislocation densities $\mathbf{b}(\tilde{t}_l,0)$ at the N collocation points. The crack-tip dislocation densities can be extrapolated from these N collocation points as:

$$\tilde{b}_i(1,0) = M_E \sum_{l=1}^{N} b_E \tilde{b}_i(\tilde{t}_l,0),$$

for the right crack tip, and

$$\tilde{b}_i(-1,0) = M_E \sum_{l=1}^{N} b_E \tilde{b}_i(\tilde{t}_{N+1-l},0),$$

for the left crack tip (Hills *et al.*, 1996), where $M_E = 1/N$, $b_E = \sin[\frac{\pi(2l-1)(2N)}{4N}]/\sin[\frac{\pi(2l-1)}{4N}]$, and $i = 1,2$.
The stress intensity factors at the crack tip $x = a$ is defined as:

$$\mathbf{K} = K_I + iK_{II} = \lim_{x_1 \to a} \sqrt{2\pi(a-x_1)}[\sigma_{22}(x_1,0) + i\sigma_{12}(x_1,0)]$$

$$= \lim_{x_1 \to a} \sqrt{2\pi(a-x_1)}\{[\mathbf{B}_{2\alpha} f'_\alpha(z_\alpha) + \bar{\mathbf{B}}_{2\alpha} \bar{f}'_\alpha(\bar{z}_\alpha)] + i[\mathbf{B}_{1\alpha} f'_\alpha(z_\alpha) + \bar{\mathbf{B}}_{1\alpha} \bar{f}'_\alpha(\bar{z}_\alpha)]\} \tag{27}$$

Omitting the involved mathematical derivation, which is analogous to the one described in Huang and Kardomateas (1999), the stress intensity factor $\mathbf{K}$ is related to the crack-tip dislocation densities as:

$$\mathbf{K}(+1) = K_I + iK_{II} = \frac{1}{4\pi}\sqrt{\pi a}\{2\mathrm{Re}\left\{\mathbf{B}_{2\alpha}\left[\mathbf{M}\mathbf{D}\tilde{\mathbf{b}}(+1) + \sum_{j=1}^{2} \mathbf{E}_{\alpha j}(+1)\right]\right\} +$$

$$+i2\mathrm{Re}\left\{\mathbf{B}_{1\alpha}\left[\mathbf{M}\mathbf{D}\tilde{\mathbf{b}}(+1) + \sum_{j=1}^{2} \mathbf{E}_{\alpha j}(+1)\right]\right\}\} \tag{28a}$$

where Re stands for the real part of a complex variable. $\mathbf{E}$ is solved from equation (10) with $\mathbf{b} = \tilde{\mathbf{b}}(+1)$. Similarly,

$$\mathbf{K}(-1) = K_I + iK_{II} = -\frac{1}{4\pi}\sqrt{\pi a}\{\mathbf{2Re}\left\{\mathbf{B}_{2\alpha}\left[\mathbf{MDb}(-\mathbf{1}) + \sum_{\mathbf{j=1}}^{2} \mathbf{E}_{\alpha\mathbf{j}}(-\mathbf{1})\right]\right\} +$$

$$i2\mathrm{Re}\left\{\mathbf{B}_{1\alpha}\left[\mathbf{MDb}(-1) + \sum_{j=1}^{2} \mathbf{E}_{\alpha j}(-1)\right]\right\}\} \tag{28b}$$

RESULTS AND DISCUSSIONS

The method presented above is implemented in computer code for both the bimaterial half plane and the bimaterial strip. To validate this method, we choose the two materials to be identical. Under this condition, the bimaterial solution should be the same as that for the homogeneous anisotropic material. It is easy for us to verify the solution on the bimaterial half plane because the analytical solution for a dislocation in a homogeneous anisotropic half plane is known (Atkinson and Eftaxiopoulos, 1991; Lee, 1990). A $90^0/90^0$ composite laminate is used in the calculation. Typical data for graphite/epoxy were used, i.e., moduli in GPa: $E_L = 130, E_T = 10.5, G_{LT} = 6$, and Poisson's ratio $\mu_{LT} = 0.28$, where L and T are the directions along and perpendicular to the fibers, respectively. First, the convergence of the numerical integration is checked and the results are listed in table 1. The free boundary is located at $y = -1$ and the dislocation $\mathbf{b} = \{1,0\}^T$ is located at $z_0 = 0$. Obviously, the stress components converge very well and agree with the analytic solution closely.

Table 2 gives the comparison of the present solution with the half plane analytic solution for stress points surrounding a dislocation $\mathbf{b} = \{1,0\}^T$ at $z_0 = 0$. Again, we obtained excellent agreements between these two solutions.

Table 1: Convergence of Stresses for a Dislocation in a Homogeneous Anisotropic Half Plane

Stresses	$z = x_1 + ix_2$	Number of Integration Points N_2								Lee's (1990) Solution
		100	200	300	400	600	800	1000	1200	
σ_{11}	0.1832	0.5403	0.5672	0.5473	0.5342	0.5321	0.5339	0.5342	0.5342	0.5346
	4.9384	0.1546	0.1552	0.1554	0.1559	0.1562	0.1562	0.1562	0.1562	0.1567
	9.6937	0.0487	0.0486	0.0491	0.0492	0.0492	0.0492	0.0492	0.0492	0.0498
σ_{12}	0.1832	5.9905	5.8304	5.7627	5.786	5.8319	5.8375	5.8368	5.8367	5.8379
	4.9384	0.0587	0.0582	0.0624	0.0638	0.0635	0.0634	0.0635	0.0635	0.0635
	9.6937	0.0078	0.0098	0.0107	0.0104	0.0102	0.0103	0.0103	0.0103	0.0103
σ_{22}	0.1832	-1.2452	-1.8569	-1.6395	-1.4415	-1.4375	-1.4799	-1.4865	1.4864	1.4870
	4.9384	-0.0166	0.0178	0.0258	0.021	0.0175	0.0179	0.0179	0.0179	0.0179
	9.6937	-0.0036	0.0075	0.0025	0.0009	0.0016	0.0016	0.0016	0.0016	0.0016

Table 2: Comparison Between the Present Method and the Analytical Solution of Lee (1990) for a Homogeneous Anisotropic Half Plane

$z = x_1 + ix_2$	Present			Lee's (1990) Solution		
	σ_{11}	σ_{12}	σ_{22}	σ_{11}	σ_{12}	σ_{22}
-0.5-0.5i	2.2035	-0.286	-0.6585	2.2046	-0.2835	-0.6724
-0.5i	2.7879	0	4.8655	2.7868	0	4.8948
0.5-0.5i	2.2035	0.286	-0.6585	2.2046	0.2835	-0.6724
0.5	0.484	1.9379	-0.2531	0.4844	1.9378	-0.2535
0.5+0.5i	-1.2243	0.3408	0.2652	-1.2239	0.3407	0.2649
0.5i	-1.7408	0	-9.1689	-1.7403	0	-9.1699
-0.5+0.5i	-1.2243	-0.3408	0.2652	-1.2239	-0.3407	0.2649
-0.5	0.484	-1.9379	-0.2531	-0.4844	-1.9378	-0.2535

Table 3: Results Comparison for Dislocations in a Homogeneous Anisotropic Strip

$z = x_1 + ix_2$	Present			Huang & Kardomateas (1999)		
	σ_{11}	σ_{12}	σ_{22}	σ_{11}	σ_{12}	σ_{22}
9.6937	0	0	0.0001	0	0	0.0001
8.9835	0	0	0	0	-0.0001	-0.0001
7.8774	0	0	-0.0001	0	0	-0.0001
6.4835	0	0	-0.0001	0	-0.0001	-0.0001
4.9384	0	0	-0.0002	0	0	-0.0002
3.3934	0.0015	0	-0.002	0.0015	0	-0.0021
1.9995	0.0494	0	-0.0629	0.0494	-0.0001	-0.0629
0.8934	0.7824	0.0001	-0.9664	0.7824	0.0002	-0.9663
0.1832	6.2163	-0.001	13.5984	6.2164	-0.0001	13.5965

The solution for dislocations in the interface with two 90^0 laminates is compared with the numerical solution presented by Huang and Kardomateas (1999) for the homogeneous anisotropic infinite strip, which superposes an array of dislocation to the homogeneous half plane solution. Table 3 gives the comparison of the results from these two numerical methods. The dislocation $\mathbf{b} = \{0, 1\}^T$ is located at $z_0 = 0$. The points chosen for the presentation of the stress data for Tables 1 and 3 are actually some of the collocation points for the numerical integration.

To study the effect of materials on the stress intensity factors(SIF) of an interface crack, we compared the interface crack in a $90^0/0^0$ laminate with a $90^0/90^0$ homogeneous material (the crack and the interface are perpendicular to the 90^0 fibers). Table 4 gives the SIFs for cracks in the half plane. The crack is located at the interface. The distance between the free boundary and the interface is 1 and the SIF is normalized as $\bar{\mathbf{K}} = \mathbf{K}/\left(T_{22}\sqrt{\pi a}\right)$, where 2a is the crack length. Since the results for the SIF are normalized with T_{22}, the unit of length used for the distance between the free boundary and the interface does not matter as long as it is consistent with the external stress T_{22}. The external load T_{22} is uniformly distributed along the crack surfaces and T_{12} is chosen to be zero.

Table 4: Normalized Stress Intensity Factors for Interface Cracks in a Bimaterial Half Plane

a	$90^0/90^0$				$90^0/0^0$			
	$\bar{K}_I(+1)$	$\bar{K}_{II}(+1)$	$\bar{K}_I(-1)$	$\bar{K}_{II}(-1)$	$\bar{K}_I(+1)$	$\bar{K}_{II}(+1)$	$\bar{K}_I(-1)$	$\bar{K}_{II}(-1)$
0.5	1.5156	0.0841	1.5156	-0.0841	1.0987	0.0071	1.0987	-0.0071
1	2.2763	0.2876	2.2763	-0.2876	1.3101	0.0392	1.3101	-0.0392
1.5	3.1465	0.5584	3.1465	-0.5584	1.5583	0.0927	1.5583	-0.0927
2	4.1138	0.8812	4.1138	-0.8812	1.8257	0.1622	1.8257	-0.1622
2.5	5.1682	1.2476	5.1682	-1.2476	2.1085	0.2446	2.1085	-0.2446
3	6.3053	1.6536	6.3053	-1.6536	2.4053	0.3381	2.4053	-0.3381
3.5	7.5197	2.0955	7.5197	-2.0955	2.7150	0.4413	2.7150	-0.4413
4	8.808	2.5707	8.808	-2.5707	3.0370	0.5534	3.0370	-0.5534
4.5	10.1697	3.0782	10.1697	-3.0782	3.3708	0.6735	3.3708	-0.6735
5	11.6194	3.6214	11.6194	-3.6214	3.7158	0.8010	3.7158	-0.8010

Table 5 gives the SIF comparison for interface cracks in an infinite strip. The thickness of both laminates is 1. The crack length and external load are the same as those used in the half plane study. The results for the infinite strip demonstrate that the material mismatch makes the crack become more of a mixed-mode type even though it causes a decrease in the mode-I stress intensity factor, K_I, while the material mismatch in the half plane make both the stress intensity factors K_I, K_{II}, and the mode mixity $\psi = \tan^{-1}(K_I/K_{II})$ decrease.

Table 5: Normalized Stress Intensity Factors for Interface Cracks in Infinite Strip

a	$90^0/90^0$				$90^0/0^0$			
	$\bar{K}_I(+1)$	$\bar{K}_{II}(+1)$	$\bar{K}_I(-1)$	$\bar{K}_{II}(-1)$	$\bar{K}_I(+1)$	$\bar{K}_{II}(+1)$	$\bar{K}_I(-1)$	$\bar{K}_{II}(-1)$
0.5	1.8205	0.0005	1.8205	-0.0005	1.3292	-0.1307	1.3292	0.1307
1	3.0626	0.0016	3.0626	-0.0016	1.8603	-0.4227	1.8603	0.4227
1.5	4.5144	0.0037	4.5144	-0.0037	2.4761	-0.796	2.4761	0.796
2	6.149	0.0073	6.149	-0.0073	3.1576	-1.2337	3.1576	1.2337
2.5	7.9429	0.013	7.9429	-0.013	3.8972	-1.7264	3.8972	1.7264
3	9.8659	0.0209	9.8659	-0.0209	4.6892	-2.2683	4.6892	2.2683
3.5	11.9311	0.0274	11.9311	-0.0274	5.5183	-2.8522	5.5183	2.8522

ACKNOWLEDGEMENTS

The financial support of the Office of Naval Research, Ship Structures S& T Division, Grant N00014-90-J-1995, and of the National Rotorcraft Technology Center through CERT, Grant NCC2-945, and the interest and encouragement of the Grant Monitors, Dr. Y.D.S. Rajapakse, Dr. G. Anderson and Dr. T.K. O'Brien are both gratefully acknowledged.

REFERENCES:

Atkinson, C. and Eftaxiopoulos, D.A. (1991) "Interaction Between a Crack and a Free or Welded Boundary in Media with Arbitrary Anisotropy", *Int. J. Fract.*, 50, p159–182.

Civelek, M.B. and Erdogan, F. (1982), "Crack Problems for a Rectangle Plate and an Infinite Strip", *Int. J. Fract.*, 19, p139 –159.

Eshelby, J. D., Read, W.T. and Cockily, W. (1953) "Anisotropic Elasticity with Applications to Dislocation Theory", *Acta Meta.*, Vol. 1, p 251 – 259.

Hills, D.A., Kelly, P.A., Di, D.N. and Korsunsky, A.M. (1996) *Solution of Crack Problems; the Distributed Dislocation Technique*, Kluwer Academic Publishers, U.K.

Huang, H. and Kardomateas, G.A. (1999) "Single-edge and Double-edge Cracks in a Fully Anisotropic Strip", in press, *Journal of Engineering Materials and Technology, ASME.*

Jagannadham, K. and Marcinkowski, M.J. (1979) "Surface Dislocation Model of a Finite Stressed Solid", *Materials Science and Engineering*, 38, p259 – 270.

Lee, J.C. (1990) "Analysis of a Fiber Bridged Crack Near a Free Surface in Ceramic·Matrix Composites", *Eng. Fract. Mech.*, Vol. 37, No. 1, p209– 219.

Qu, J. and Li, Q. (1991) "Interfacial Dislocation and Its Applications to Interface Cracks in Anisotrpic Bimaterials", *J. Elast.*, 26, p169–195.

Stroh, A.N. (1958) "Dislocations and Cracks in Anisotropic Elasticity", *Philos. Mag.*, 7, p625 – 646.

Suo, Z. (1990) "Delamination Specimens for Orthotropic Materials", *J. Appl. Mech.*, Vol. 57, p. 627 – 634.

Suo, Z. and Hutchison, J.W. (1990) "Interface Crack Between Two Elastic Layers", *Int. J. Fracture*, 43, p1–18.

Ting, T. (1986) "Explicit Solution and Invariance of the Singularities at an Interface Crack in Anisotropic Composites", *Int. J. Solids Structures*, Vol. 22, No. 9, p965 – 983.

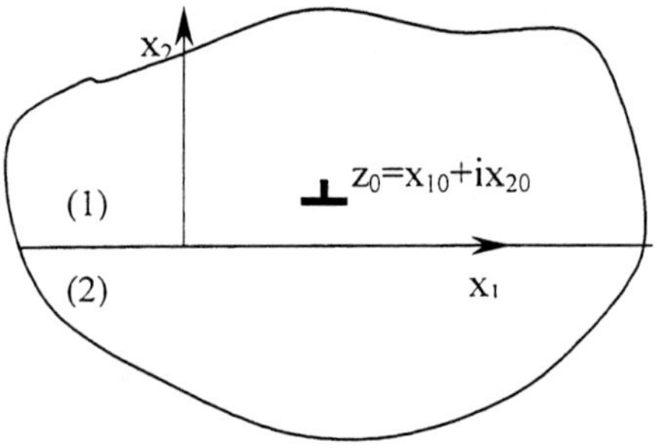

Figure 1: Dislocation in Bi-material Infinite Plane

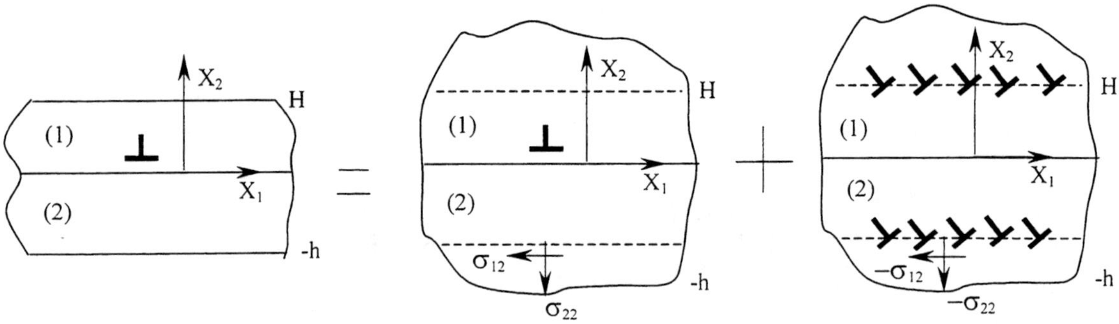

Figure 2: Decomposition of a Dislocation in Bi-material Thin Strip

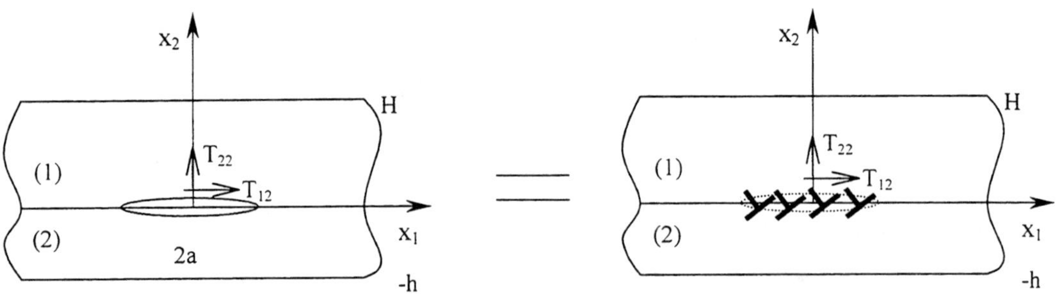

Figure 3: Interface Crack in Bi-material Infinite Strip

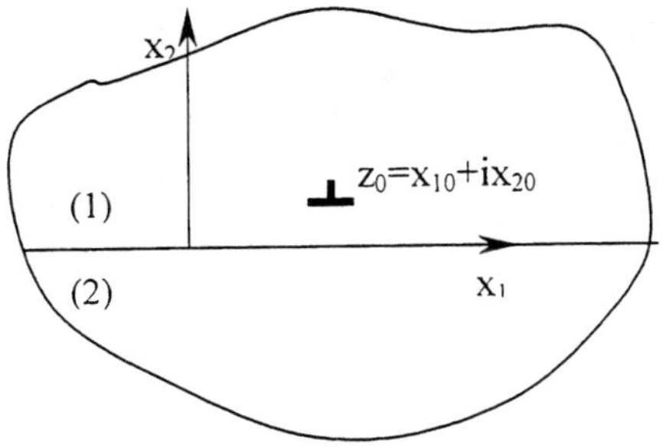

Figure 1: Dislocation in Bi-material Infinite Plane

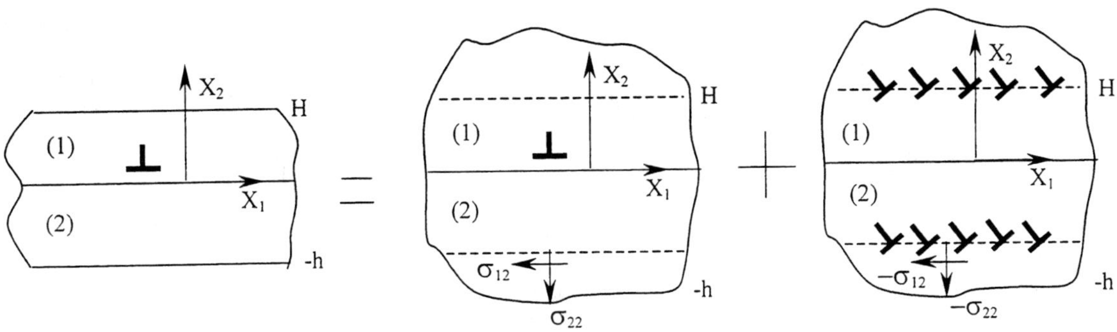

Figure 2: Decomposition of a Dislocation in Bi-material Thin Strip

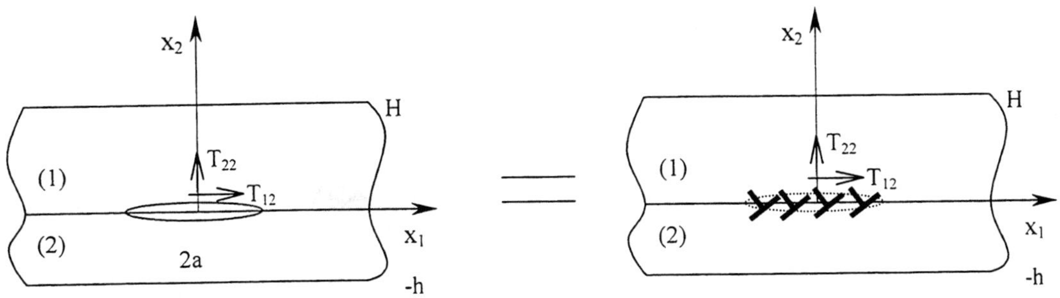

Figure 3: Interface Crack in Bi-material Infinite Strip

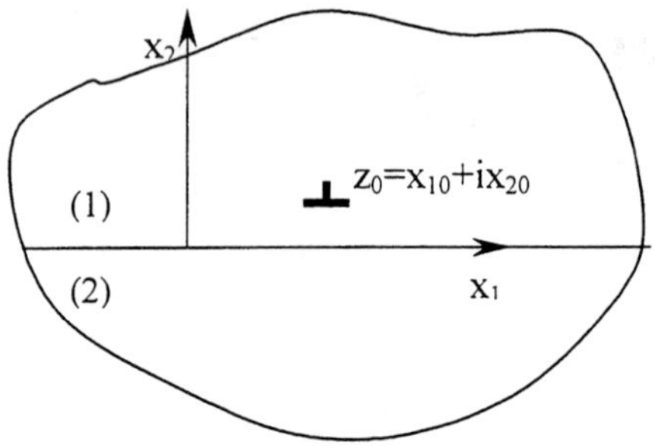

Figure 1: Dislocation in Bi-material Infinite Plane

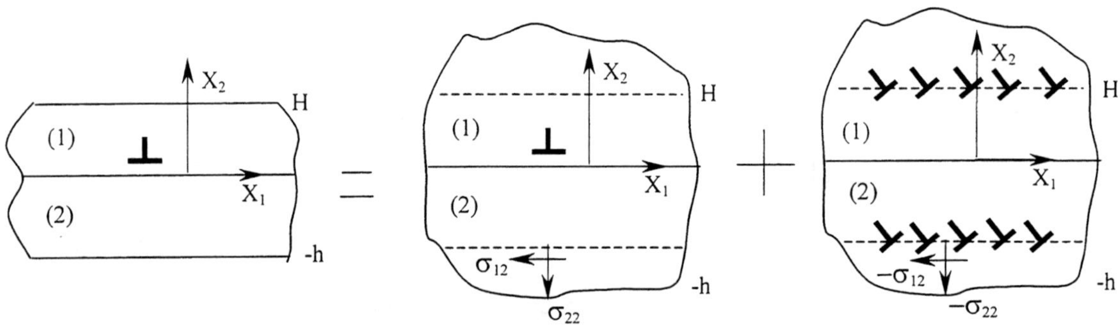

Figure 2: Decomposition of a Dislocation in Bi-material Thin Strip

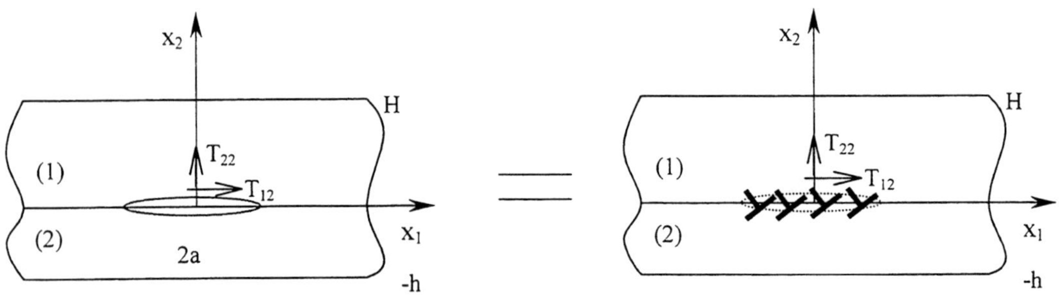

Figure 3: Interface Crack in Bi-material Infinite Strip

AMD-Vol. 235, Thick Composites for Load Bearing Structures
ASME 1999

MIXED MODE STRESS SINGULARITIES IN ANISOTROPIC COMPOSITES

Wan-Lee Yin
Georgia Institute of Technology
Atlanta, GA 30332-0355

ABSTRACT

Multi-material wedges composed of fully anisotropic elastic sectors generally show intrinsic coupling of the anti-plane and in-plane modes of deformation. Each anisotropic sector has three complex conjugate pairs of material eigensolutions whose form of expression depends on five distinct types of anisotropic materials. Continuity of the displacements and the tractions across the sector interfaces and the traction-free conditions on two exterior boundary edges determine an infinite sequence of eigenvalues and eigensolutions of the multi-material wedge. These eigensolutions are linearly combined to match the traction-boundary data (generated by global finite element analysis of the structure) on a circular path encircling the singularity. The analysis method is applied to a bimaterial wedge near the free edge of a four-layer angle-ply laminate, and to a trimaterial wedge surrounding the tip of an embedded oblique crack in a three-layer composite. Under a uniform temperature load, the elasticity solution based on the eigenseries yields interfacial stresses that are significantly different from the asymptotic solution (given by the first term of the eigenseries), even as the distance from the singularity decreases to subatomic scales. Similar observations have been found previously for isotropic and orthotropic multi-material wedges. This raises serious questions with regard to characterizing the criticality of stress singularity exclusively in terms of the asymptotic solution and the associated stress intensity factors or generalized stress intensity factors.

1. INTRODUCTION

When a composite structure is subjected to mechanical or hygrothermal loading, intense localized stress often arise in critical regions near free edges, sharp corners, crack tips, joints and interfaces due to the mismatch in the elastic and hygrothermal properties of dissimilar materials. The elasticity solutions of problems of this type generally involve a singular stress field at the point of intersection of interfaces and free edges. The character of singularity is determined by the local geometry of the multi-material wedge surrounding the singularity and the anisotropic elastic constants of the various sectors of the wedge. Such singular solutions cannot be obtained by conventional finite element analysis. Although special singular elements incorporating the behavior of the dominant eigensolution have been used in the study of free-edge effects and interface crack problems, the accuracy of the results was found to depend appreciably on the size of the singular element. Furthermore, the singular element used in previous studies often contain the dominant eigensolution only, without attempting to account for the effects of higher eigensolutions.

It is well known that the region of intense stress is localized in a small vicinity of the singularity. The effects of the singularity decay rapidly away from the vertex of the wedge. Hence the stress solution of a conventional finite element analysis using non-singular elements yields accurate results in regions away from the singularities. A two-step analysis approach may be adopted by using the results of a conventional finite element analysis to provide the boundary traction data on the circular boundary of a bimaterial or multimaterial wedge. An elasticity analysis of the wedge is then performed to determine its eigenvalues and eigensolutions, and the eigensolutions are linearly combined with appropriate coefficients to fit the traction data on the circular boundary of the wedge.

This two-step approach has been applied to multimaterial wedges composed of isotropic and orthotropic sectors (Yin, 1996; 1997). It was found that the eigenseries converges rapidly and only a small number of leading terms contribute significantly to the stress field near the singularity. For the problem of an interface crack between two dissimilar materials, the elasticity solution based on an eigenseries yields an energy release rate that differs by less than one percent from the value of the J-integral along a remote path (Yin, 1998). This and other comparisons confirm the accurate of the two-step substructure approach.

Multimaterial wedges composed of isotropic and orthotropic sectors show no coupling between in-plane and anti-plane deformations. Hence all eigensolutions are either plane deformation modes or anti-plane modes. In contrast, if the material of one or more sectors is fully anisotropic, or is orthotropic but the out-of-plane *z*-direction is not an orthotropic axis (as in the case of an angle-ply layer in a laminated composite), then the in-plane and anti-plane deformations are intrinsically coupled in *all* static solutions of the wedge, including all eigensolutions. This general case will be investigated in the present analysis.

In the previous studies of multimaterial wedges (Yin, 1997; Yin *et al.*, 1997), isotropic and transversely isotropic sectors require special treatment because, from the viewpoint of anisotropic elasticity, they are made of *degenerate* materials which have special representations for the material eigensolutions. In the case of fully anisotropic sectors, the material may be degenerate and may even be extradegenerate (Ting, 1996). It was shown recently that anisotropic elastic materials can be classified into five types. Explicit representations of the material eigensolutions for all five types of anisotropic materials have been found (Yin, 1999a,b). These representations provide the basis for a comprehensive analysis of multimaterial wedges with fully anisotropic sectors.

Some basic equations of anisotropic elasticity are given in Sec. 2. A general formulation of the eigenvalue problem for a multimaterial wedge is shown in Sec. 3. Section 4 deals with a bimaterial wedge near the free edge of a four-layer angle-ply laminate subjected to a uniform temperature load. In Sec. 5 we investigate a tri-sector wedge surrounding the lower tip of an oblique crack in the middle layer of a three-layer laminate subjected also to $\Delta T = -1°$. The last two sections also contain some discussion and interpretations of the results.

2. TWO-DIMENSIONAL ANISOTROPIC ELASTICITY

Let α_{ij} (i, j = 1, . . . , 6) denote the anisotropic elastic compliance constants relating the strain components $\epsilon_x, \epsilon_y, \epsilon_z, \gamma_{yz}, \gamma_{xz}, \gamma_{xy}$ to the stress components $\sigma_x, \sigma_y, \sigma_z, \tau_{yz}, \tau_{xz}, \tau_{xy}$, and let

$$\beta_{ij} = \alpha_{ij} - \alpha_{i3}\,\alpha_{j3}/\alpha_{33} \qquad \text{(for i, j} \neq 3).$$

Then, for generalized plane deformations, one has

$$\{\epsilon\} = [\beta]\,\{\sigma\}, \tag{2.1}$$

where $\{\epsilon\} = \{\epsilon_x, \epsilon_y, \gamma_{yz}, \gamma_{xz}, \gamma_{xy}\}^T$, $\{\sigma\} = \{\sigma_x, \sigma_y, \tau_{yz}, \tau_{xz}, \tau_{xy}\}^T$ and

$$[\beta] = \begin{bmatrix} \beta_{11} & \beta_{12} & \beta_{14} & \beta_{15} & \beta_{16} \\ \beta_{12} & \beta_{22} & \beta_{24} & \beta_{25} & \beta_{26} \\ \beta_{14} & \beta_{24} & \beta_{44} & \beta_{45} & \beta_{46} \\ \beta_{15} & \beta_{25} & \beta_{45} & \beta_{55} & \beta_{56} \\ \beta_{16} & \beta_{26} & \beta_{46} & \beta_{56} & \beta_{66} \end{bmatrix}$$

In the absence of body forces, the equilibrium conditions imply that $\{\sigma\}$ may be represented by the derivatives of a pair of stress functions $F(x, y)$ and $\Psi(x, y)$:

$$\sigma_x = F,_{yy}, \qquad \sigma_y = F,_{xx}, \qquad \tau_{xy} = -F,_{xy},$$
$$\tau_{xz} = \Psi,_y, \qquad \tau_{yz} = -\Psi,_x, \tag{2.2}$$

We seek solutions for the displacements $\mathbf{u} \equiv \{u, v, w\}$ and the stress potentials $\mathbf{q} \equiv \{F,_y, -F,_x, \Psi\}$ of the following form

$$\mathbf{u} = \mathbf{a}\, f(x + \mu y), \qquad \mathbf{q} = \sum \mathbf{b}\, f(x + \mu y) \tag{2.3a,b}$$

where f is an arbitrary function. The complex parameter μ and the complex vectors **a** and **b** will be later identified as material eigenvalues and eigenvectors. The strain and stress field associated with Eqs. (2.3a,b) are easily obtained by differentiation. Since

$$\tau_{xy} = -\partial_x F,_y = -\sum b_1 f'(x + \mu y) = \partial_y(-F,_x) = \sum b_2\, \mu\, f'(x + \mu_i y)$$

one has

$$b_1 = -\mu\, b_2 \tag{2.4}$$

Equations (2.1) and (2.4) yield

$$\mathbf{E}(\mu)\,\mathbf{a} = [\beta]\,\mathbf{P}(\mu) \begin{Bmatrix} b_2 \\ b_3 \end{Bmatrix} \tag{2.5}$$

Pre-multiplication of the last equation by $\mathbf{E}^T(\mu)\,[\beta]^{-1}$ and $\mathbf{P}(\mu)^T$ yield, respectively,

$$\mathbf{E}^T(\mu)\,[\beta]^{-1}\,\mathbf{E}(\mu)\,\mathbf{a} \equiv \Gamma(\mu)\,\mathbf{a} = \mathbf{0}, \tag{2.6}$$

$$\mathbf{P}^T(\mu)\,[\beta]\,\mathbf{P}(\mu) \begin{Bmatrix} b_2 \\ b_3 \end{Bmatrix} = \mathbf{M}(\mu) \begin{Bmatrix} b_2 \\ b_3 \end{Bmatrix} = \mathbf{0}. \tag{2.7}$$

where the matrix functions $\mathbf{E}(\mu)$ and $\mathbf{P}(\mu)$ are defined by

$$\mathbf{E}(\mu) = \begin{bmatrix} 1 & 0 & 0 \\ 0 & \mu & 0 \\ 0 & 0 & \mu \\ 0 & 0 & 1 \\ \mu & 1 & 0 \end{bmatrix} \qquad \mathbf{P}(\mu) = \begin{bmatrix} -\mu^2 & 0 \\ -1 & 0 \\ 0 & -1 \\ 0 & \mu \\ \mu & 0 \end{bmatrix} \tag{2.8a,b}$$

Equation (2.7)has a nontrivial solution for **a** if and only if μ is a root of the characteristic equation

$$\Delta(\mu) \equiv |\Gamma(\mu)| = 0\,. \tag{2.9}$$

Similarly, Eq. (2.6) has a nontrivial solution for $\{b_2, b_3\}$ if and only if μ satisfies

$$\delta(\mu) \equiv |\mathbf{M}(\mu)| = 0\,. \tag{2.10}$$

The two conditions are in fact equivalent, and Eqs. (2.9) and (2.10) yield three identical pairs of complex conjugate roots. These roots are the material eigenvalues of the anisotropic material. For each eigenvalue μ, the associated **a**-vector may be solved from Eq. (2.6) and the **b**-vector may be obtained from Eqs. (2.7) and (2.4). Substituting **a**, **b** and μ into Eqs. (2.3a,b), we obtain an eigensolution of the material associated with the material eigenvalue μ.

It is easily seen that the complex conjugate of the eigenvalue μ is associated with vectors that are the complex conjugates of **a** and **b**. If the characteristic equation (2.10) has three distinct pairs of complex conjugate roots, we let **A** denote the 3×3 matrix whose column are the three **a**-vectors associated with the eigenvalues that have *positive* imaginary parts. We let **B** denote the corresponding matrix of **b**-vectors. Then the 6×6 matrix

$$\mathbf{H} \equiv \begin{bmatrix} \mathbf{A} & \bar{\mathbf{A}} \\ \mathbf{B} & \bar{\mathbf{B}} \end{bmatrix} \tag{2.11}$$

is the matrix composed of six-dimensional eigenvectors. Here the overbars denote complex conjugates. In the present study, the arbitrary function f of Eq. (2.3) has the form $f(z) = z^\lambda$, where λ may be real or complex.

If the characteristic equation has a repeated root μ and the number of independent **a**- or **b**- vectors associated with μ is smaller than the multiplicity of μ, then the material is called degenerate or extra-degenerate, depending on whether the difference between the two numbers is 1 or 2. In such cases, the set of independent eigenvectors must be supplemented by "generalized eigenvectors". The later may be obtained by the "derivative rule" (Yin, 1999a,b). The matrix **H** of Eq. (2.11) is then composed of eigenvectors and generalized eigenvectors. The eigensolutions of the displacements and the stress potentials corresponding to a generalized eigenvector are *not* given by Eqs. (2.3a,b). The correct form may also be found in Yin (1999a,b). The degenerate case often arises because isotropic materials and transversely isotropic material are degenerate.

3. A MULTI-MATERIAL WEDGE

A multi-material wedge is composed of *N* consecutive sectors of isotropic or anisotropic materials. All interfaces of the sectors converge at the vertex of the wedge. The *k*th interface, $\theta = \theta_k$, separates the *k*th sector from the (*k+1*)th sector. The first and the *N*th sectors are adjacent, respectively, to traction-free exterior boundary edges $\theta = \theta_0$ and $\theta = \theta_N$.

If the material of the kth sector is non-degenerate, we define the 6×6 diagonal matrix

$$\Theta^{(k)}(\theta) = <(\cos\theta + \mu_1^{(k)}\sin\theta)^\lambda,\ (\cos\theta + \mu_2^{(k)}\sin\theta)^\lambda,\ (\cos\theta + \mu_3^{(k)}\sin\theta)^\lambda,$$

$$(\cos\theta + \bar{\mu}_1^{(k)}\sin\theta)^\lambda,\ (\cos\theta + \bar{\mu}_2^{(k)}\sin\theta)^\lambda,\ (\cos\theta + \bar{\mu}_3^{(k)}\sin\theta)^\lambda> \qquad (3.1)$$

For a degenerate sector with $\mu_1 = \mu_2 \neq \mu_3$, or an extradegenerate material with $\mu_1 = \mu_2 = \mu_3$, we let

$$\Theta^{(k)}(\theta) = \begin{bmatrix} \Theta_1 & \mathbf{0} \\ \mathbf{0} & \Theta_2 \end{bmatrix} \qquad (3.2)$$

where, for the degenerate case,

$$\Theta_1 = \begin{bmatrix} (\cos\theta + \mu_1^{(k)}\sin\theta)^\lambda & \lambda\sin\theta\,(\cos\theta + \mu_1^{(k)}\sin\theta)^{\lambda-1} & 0 \\ 0 & (\cos\theta + \mu_1^{(k)}\sin\theta)^\lambda & 0 \\ 0 & 0 & (\cos\theta + \mu_3^{(k)}\sin\theta)^\lambda \end{bmatrix} \qquad (3.3)$$

whereas for the extradegenerate case

$$\Theta_1 = \begin{bmatrix} (\cos\theta+\mu_1^{(k)}\sin\theta)^\lambda & \lambda\sin\theta\,(\cos\theta+\mu_1^{(k)}\sin\theta)^{\lambda-1} & \lambda^2\sin^2\theta\,(\cos\theta+\mu_1^{(k)}\sin\theta)^{\lambda-2} \\ 0 & (\cos\theta+\mu_1^{(k)}\sin\theta)^\lambda & 2\lambda\sin^2\theta\,(\cos\theta+\mu_1^{(k)}\sin\theta)^{\lambda-1} \\ 0 & 0 & (\cos\theta + \mu_1^{(k)}\sin\theta)^\lambda \end{bmatrix}$$

$$(3.4)$$

In both cases the matrix Θ_2 is obtained from Θ_1 by merely replacing $\mu_1^{(k)}$ and $\mu_3^{(k)}$, respectively, by their complex conjugates $\bar{\mu}_1^{(k)}$ and $\bar{\mu}_3^{(k)}$ (but keeping λ unchanged).

In the kth sector, the solution for the displacements and stress potentials $\mathbf{h}^{(k)} \equiv \{u, v, w, F_{,y}, -F_{,x}, \Psi\}$, associated with the complex parameter λ, is given by

$$\mathbf{h}^{(k)}(r, \theta) = r^\lambda\, \mathbf{H}^{(k)}\, \Theta^{(k)}(\theta)\, \mathbf{c}^{(k)}, \qquad (3.5)$$

where the components of the 6-vector $\mathbf{c}^{(k)}$ are the coefficients combining the six independent solutions of the sector. The matrix $\mathbf{H}$, defined by Eq. (2.11), are composed of the eigenvectors of the sector.

Continuity of the displacements and the tractions on the kth interface between the kth and $(k+1)$th sectors requires

$$\mathbf{H}^{(k+1)}\, \Theta^{(k+1)}(\theta_k)\, \mathbf{c}^{(k+1)} = \mathbf{H}^{(k)}\, \Theta^{(k)}(\theta_k)\, \mathbf{c}^{(k)}. \qquad (3.6)$$

This implies the recurrence relation for the vectors $\mathbf{c}^{(k)}$:

$$\mathbf{c}^{(k+1)} = \mathbf{T}_k\, \mathbf{c}^{(k)}, \qquad \text{where} \qquad \mathbf{T}_k \equiv [\Theta^{(k+1)}(\theta_k)]^{-1}\,[\mathbf{H}^{(k+1)}]^{-1}\, \mathbf{H}^{(k)}\, \Theta^{(k)}(\theta_k). \qquad (3.7)$$

Traction-free conditions on the two exterior boundary edges $\theta = \theta_0$ and $\theta = \theta_N$ imply

$$< 0, 0, 0, 1, 1, 1 > \quad \mathbf{H}^{(1)} \Theta^{(1)}(\theta_0)\, \mathbf{c}^{(1)} = \mathbf{0}\,, \tag{3.8a}$$

$$< 0, 0, 0, 1, 1, 1 > \quad \mathbf{H}^{(N)} \Theta^{(N)}(\theta_N)\, \mathbf{c}^{(N)} = \mathbf{0}\,. \tag{3.8b}$$

If one or both boundary edges are rigidly fixed instead of free, then the diagonal matrix < 0, 0, 0, 1, 1, 1> in the corresponding edge conditions should be replaced by < 1, 1, 1, 0, 0, 0 >.

The recurrence relation implies

$$\mathbf{c}^{(N)} = \mathbf{T}_{N-1}\, \mathbf{T}_{N-2} \cdots \mathbf{T}_1\, \mathbf{c}^{(1)} \tag{3.9}$$

Substitution into the boundary conditions on the edge $\theta = \theta_N$, and combining the results with the boundary conditions on $\theta = \theta_0$, one obtains a system of six homogenous linear equations for the six components of $\mathbf{c}^{(1)}$. The coefficients of the equations are algebraic and exponential functions of λ with multiple exponents. The system has a nontrivial solution $\mathbf{c}^{(1)}$ if the determinant of the matrix of coefficients vanishes. This yields a characteristic equation for the complex parameter λ. The roots of the equations are the eigenvalues of the multimaterial wedge. Generally the roots must be found by numerical methods.

Once an eigenvalue λ is found, the solutions $\mathbf{c}^{(1)}$ of the six homogeneous linear equations may be determined up a complex scalar factor. Then all other vectors $\mathbf{c}^{(k)}$ $(k = 2, 3, \ldots, N)$ may be calculated using the recurrence relation, and Eq. (3.5) gives the displacements and the stress potentials in each sector. They provide an *eigensolution* of the multimaterial wedge corresponding to the eigenvalue λ. If the real part of λ is smaller than one, then the stress field near $r = 0$ has the $r^{-(1-\lambda)}$ type singularity. Such eigenvalues are called singular eigenvalues. The roots of the characteristic equation with the real parts smaller or equal to zero are ignored because the associated eigensolution requires infinite strain energy.

If λ is a repeated root of the characteristic equation, then Eq. (3.8) may give more than one independent solution vectors $\mathbf{c}^{(1)}$. Each vector yields an independent eigensolution of the multimaterial wedge base on the recurrence relation (3.7) and Eq. (3.5). However, the number of independent solution vectors of Eq. (3.8) may be smaller than the multiplicity of the eigenvalue λ. In such cases, additional eigensolutions involving $log(x + \mu_i y)$ factors may be found from an equation for $d\mathbf{c}^{(k)}/d\lambda$ that can be obtained by differentiating Eq. (3.8) with respect to λ.

For the bimaterial wedge associated with an interface crack, we let $\theta_0 = -\pi$ and $\theta_2 = \pi$ be the crack faces and $\theta_1 = 0$ be the intact interface. Since $\Theta^{(1)}(0) = \Theta^{(2)}(0) = \mathbf{I}$, the displacements and interlaminar stresses on the interface are given by

$$\mathbf{h}\,(r, 0) = r^{\lambda}\, \mathbf{H}^{(1)}\, \mathbf{c}^{(1)} = r^{\lambda}\, \mathbf{H}^{(1)}\, \mathbf{c}^{(1)}\,, \tag{3.10}$$

Traction-free conditions on the two edges reduce to exceptionally simple forms

$$< 0, 0, 0, 1, 1, 1 > \mathbf{H}^{(1)} < e^{-i\pi\lambda}, e^{-i\pi\lambda}, e^{-i\pi\lambda}, e^{i\pi\lambda}, e^{i\pi\lambda}, e^{i\pi\lambda} > [\,\mathbf{H}^{(1)}\,]^{-1}\, \mathbf{h}(1, 0) = \mathbf{0}, \tag{3.11a}$$

$$< 0, 0, 0, 1, 1, 1 > \mathbf{H}^{(2)} < e^{i\pi\lambda}, e^{i\pi\lambda}, e^{i\pi\lambda}\, e^{-i\pi\lambda}, e^{-i\pi\lambda}, e^{-i\pi\lambda} > [\,\mathbf{H}^{(2)}\,]^{-1}\, \mathbf{h}(1, 0) = \mathbf{0}. \tag{3.11b}$$

The simplicity is extraordinary because, when θ_0, θ_1 and θ_2 assume the angles $-\pi$, 0 and π, respectively, the values of the matrices $\Theta^{(1)}(\theta)$ and $\Theta^{(2)}(\theta)$ for these angles depend only on the parameter λ and *not* on the sector eigenvalues $\mu_1^{(1)}$, $\mu_2^{(1)}$, $\mu_3^{(1)}$, $\bar{\mu}_1^{(1)}$, $\bar{\mu}_2^{(1)}$, $\bar{\mu}_3^{(1)}$, $\mu_1^{(2)}$, $\mu_2^{(2)}$, $\mu_3^{(2)}$, $\bar{\mu}_1^{(2)}$, $\bar{\mu}_2^{(2)}$ and $\bar{\mu}_3^{(2)}$. *Regardless of material degeneracy*, Eqs. (3.11a,b) lead to an exceptionally simple characteristic equation with constant coefficients in which the unknown parameter λ appears in the form $e^{2i\pi\lambda}$. It may be simplified to the following form

$$(\sin \pi\lambda) \cdot \text{Determinant}\,[\,\text{Re}[\mathbf{A}^{(1)}(\mathbf{B}^{(1)})^{-1} - \mathbf{A}^{(2)}(\mathbf{B}^{(2)})^{-1}]$$

$$- (\text{ctn}\,\pi\lambda)\,\text{Im}[\mathbf{A}^{(1)}(\mathbf{B}^{(1)})^{-1} + \mathbf{A}^{(2)}(\mathbf{B}^{(2)})^{-1}]\,] = 0, \qquad (3.12)$$

The first factor of the characteristic equation yields a sequence of real eigenvalues $\pi\lambda = 1, 2, 3, \ldots$ The second factor is a sextic equation for ctn $\pi\lambda$. Hence $\pi\lambda$ can be obtained from the six roots of the sextic equations. These roots are either real or form complex conjugate pairs. Therefore the complex eigenvalues λ also form complex conjugate pairs.

4. A BIMATERIAL WEDGE NEAR THE FREE EDGE OF A LAMINATE

With the exception of the interface crack problem, all other bimaterial or multi-material wedges contain sectors which are not half planes. Then $\Theta^{(k)}$ depends on $\mu_i^{(k)}$ and has different forms of expression depending on material degeneracy. The characteristic equation becomes much more complicated.

One example is the bimaterial wedge surroundign the intersection of an interface, $\theta_1 = 0$, with an orthogonal free edge whose lower and upper segments are $\theta_0 = -\pi/2$ and $\theta_2 = \pi/2$, respectively. $\Theta^{(k)}(0) = \Theta^{(k)}(0) = \mathbf{I}$ is still valid regardless of material degeneracy. If materials of both sectors are non-degenerate, then $\Theta^{(1)}(-\pi/2)$ and $\Theta^{(2)}(\pi/2)$ are diagonal matrices, *viz.*

$$\Theta^{(1)}(-\pi/2) = \langle (-\mu_1^{(1)})^\lambda,\ (-\mu_2^{(1)})^\lambda,\ (-\mu_3^{(1)})^\lambda,\ (-\bar{\mu}_1^{(1)})^\lambda,\ (-\bar{\mu}_2^{(1)})^\lambda,\ (-\bar{\mu}_3^{(1)})^\lambda \rangle, \qquad (4.1)$$

$$\Theta^{(2)}(\pi/2) = \langle (\mu_1^{(2)})^\lambda,\ (\mu_2^{(2)})^\lambda,\ (\mu_3^{(2)})^\lambda,\ (\bar{\mu}_1^{(2)})^\lambda,\ (\bar{\mu}_2^{(2)})^\lambda,\ (\bar{\mu}_3^{(2)})^\lambda \rangle, \qquad (4.2)$$

However, if one or both sectors are made of degenerate or extradegenerate materials, then off-diagonal terms appear in the corresponding expressions. Appropriate expressions are given by Eqs. (3.3) and (3.4), when evaluated at $\theta = \pm\pi/2$.

We consider the free-edge thermal stress in a four-layer laminated strip of graphite epoxy with the ply configuration $[45/-45]_S$ subjected to a 1° uniform temperature decrement. The strip is infinitely long in the *z*-direction and has the width $16h$ in the *x*-direction. All four plies have the same thickness *h*. For this laminate, Wang and Crossman (1977a,b) first performed finite element analysis of the interlaminar stresses using regular finite elements. Stress singularity was noticed but not investigated mathematically in their paper. The eigensolutions of anisotropic elasticity theory for the bimaterial wedge at the free edge, and the full elasticity solution for the strip subjected to unit axial strain load along the *z*-direction, were determined by Wang and Choi (1982). The problem has been reexamined in many subsequent studies.

Using the same elastic and thermal expansion properties of the unidirectional composite as given by these authors, we obtain eigenvalues that are close to those given by Wang and Choi (1982). It should be mentioned that 1 is a double root of the characteristic equation while all other positive integers are single roots. The double eigenvalue $\lambda = 1$ has two independent eigensolutions determined by the corresponding vectors $\mathbf{h}(1, 0)$ of Eq. (3.10). The first eigensolution, for which $\mathbf{h}(1, 0) = \{0, 1, 0, 0, 0, 0\}^T$, yields no interlaminar stresses on the interface. The second eigensolution, for which $\mathbf{h}(1, 0) = \{0, 0, 0, 0, 0, 1\}^T$, yields only the out-of-plane interfacial stress component τ_{yz}. There is only one non-integer real eigenvalue which is also the only singular eigenvalue, $\lambda = 0.974062$, for which $\mathbf{h}(1, 0) = \{-0.021804, 0, 0, 0, -0.462123, 1\}^T$. This eigensolution yields interfacial stresses σ_y and τ_{yz} but no shearing stress τ_{xy}. τ_{xy} is only produced by the eigensolutions associated with some complex eigenvalues, i.e., $1.872605 \pm 0.254448i$, $3.33179 \pm 1.136085i$, $4.94999 \pm 2.04562i$, etc. These eigensolutions have vanishing σ_y and τ_{yz} on the interface. Another group of complex eigenvalues including $2.508668 \pm 0.797126i$ and $4.092877 \pm 1.740873i$ have eigensolutions similar to the singular eigensolution, i.e., the second through the fourth elements of the associated vector $\mathbf{h}(1, 0)$ vanish.

Due to the presence of the factor r^λ on the right hand of side of Eq. (3.5), eigensolutions with Re[λ] significantly larger than 1 have negligible effects on the stress field in the immediate vicinity of the edge vertex. However, some of such eigensolutions must still be obtained and used in boundary collocation for accurate determination of the coefficients of the leading eigensolutions, because all such eigensolutions have comparable magnitudes on the collocation path $r = 1$. Past experiences suggest reasonably fast convergence of the eigenseries, i.e., in most known examples the magnitudes of the complex coefficients of the successive eigensolutions (arranged in the order of increasing Re[λ]) rapidly decrease, so that only a moderate number of leading terms need be used to approximate the elasticity solution. In this analysis of the free edge problem, we use a total of 17 eigensolutions corresponding to all eigenvalues with Re[λ]$\leq$ 5. The collocation data are generated on the path $r = h/2 = 1$ by the finite element code *PDEase* using triangular elements. These data for the normal and shearing stresses in the polar coordinates are integrated to obtain the data for $F,_r$, $F,_\theta$ and Ψ. It is preferable to use $F,_r$, $F,_\theta$ and Ψ rather than the tractions σ_r, $\tau_{r\theta}$ and $\tau_{\theta z}$ as the collocation data because the stress potentials are all continuous whereas σ_r is discontinuous across the interface.

However, the data of $F,_r$, $F,_\theta$ and Ψ based on the finite element solution must be further modified before they can be used for collocation. This is because the finite element solution of the model under the temperature load $\Delta T = -1°$ contains thermal strain, whereas the eigensolutions of the multimaterial wedge are without thermal strain. To resolve the difficulty, we first consider *free* thermal expansion (except for the reaction stress σ_z to the imposed condition $\epsilon_z = 0$) of the four separate layers under $\Delta T = 1°$, which results in vanishing stresses in all layers but yields incompatible strains between the 45° and $-45°$ layers. In the present case the incompatible strains of free expansion are associated only with γ_{xz}, which has the value -3.79437×10^{-6} for the 45° layer and the opposite value for the $-45°$ layer. This differential thermal strain of the two layers may be compensated by a simple particular solution corresponding to a uniform tension stress in all four layers (also separate from one another) in the thickness direction, $\sigma_y = 3.79437/\beta_{24} = 61.7015$ psi. Under the imposed condition $\epsilon_z = 0$, the uniform stress σ_y causes no incompatibility of the tangential strain ϵ_x across the interface of the ±45° layers. When the finite element solution is combined with the two complementary solutions -- one representing free thermal expansion of the separate layers under $\Delta T = 1°$, and the other the uniform tension stress of the same layers in the thickness direction -- a solution with zero temperature load and with no interfacial discontinuity of tangential strains is obtained. It is this combined solution that is used as the data for collocation. Since the collocation data on $r = 1$ includes the contribution from the second complementary solution associated with $\sigma_y = 61.7015$ psi, an opposite uniform stress state $\sigma_y = -61.7015$ psi must be superimposed on the eigenseries to obtain the elasticity solution in the wedge.

The coefficient of the first eigensolution associated with the double root $\lambda = 1$ was found to be zero. The second associated eigensolution has the coefficient -123.326, whereas the eigensolution corresponding to the only singular eigenvalue, $\lambda = 0.974062$, has the coefficient 134.521. The coefficients of the remaining eigensolutions are relatively small.

The finite element analysis used to provide the collocation data also yields interlaminar stresses on the interface. Since the finite element analysis used only regular elements, the results cannot be accurate near the singularity. However, they are expected to be approximately valid away from the singularity. The results are shown in Fig. 1, and compared with the elasticity solution obtained by collocation. The two solutions are in agreement away from the free edge. The previous finite element solution of Wang and Crossman (1977b) for the same problem gave similar results for the interlaminar stresses. Notice that the collocation process uses only the finite element results on the collocation path $r = 1$. Yet it yields an elasticity solution whose interlaminar stresses in the interior segment $0.5 < r < 1$ agree closely with the same finite element solution.

The singular eigensolution (asymptotic solution) has a singularity of the type $r^{-0.025939}$. The behavior of the elasticity solution near $r = 0$ is best examined by normalizing the interlaminar stresses with respect to $r^{-0.025939}$, and plotting the normalized results versus r with the latter shown in the logarithmic scale. Figure 3 shows that σ_y and τ_{xy} approach the asymptotic solution (the dashed horizontal lines) rapidly as r decreases, and become almost indistinguishable from the asymptotic solutions when r diminishes to 10^{-3}. In contrast, the most prominent interlaminar stress τ_{yz} of the elasticity solution is significantly different from the asymptotic solution even as r diminishes to physically meaningless minute scales (Fig. 4). This happens because the two lowest eigenvalues are very close and the coefficients of their corresponding eigensolutions are comparable. If the asymptotic solution and the associated generalized stress intensity factors were used as the only basis for evaluating the stress criticality in the wedge and for predicting failure, it would yield very misleading results and uncertain conclusion because, for the physically meaningful range of r from about 10^{-8} to 1, the true solution of the most prominent interlaminar stress τ_{yz} is only a fraction of the value given by the asymptotic solution.

It is found that, inside the collocation path $r = 1$, τ_{yz} as determined by the seventeen-term eigenseries are nearly indistinguishable from the sum of the first three terms in the series, i.e., the singular eigensolution and the two eigensolutions associated with the double eigenvalue $\lambda = 1$ (since, as mentioned earlier, one of these has vanishing τ_{yz}, the set of three actually reduces to two). This is seen from the solid and dashed curves in Fig. 4, where the solid curve indicates the 17-term results. They are indistinguishable except in a close vicinity of $r = 1$. In contrast, partial sums of the 17-term eigenseries including 1, 3, 5, 7, 9 and 11 terms show significantly different results of σ_y and τ_{xy} in the range of r from 10^{-2} to 1. This shows that, as the number of terms in the truncated eigenseries increases, the three interfacial stresses may show distinctly different rates of convergence to their exact results. Part of the explanation may be attributed to particular eigensolutions dominated by a specific mode of shearing or peeling action across one or other interfaces.

5. A TRI-SECTOR WEDGE AROUND THE TIP OF AN OBLIQUE MATRIX CRACK

We next consider a tri-sector wedge surrounding the tip of an oblique crack in a 0° unidirectional layer which is sandwiched between two identical but differently oriented layers. All three layers have the same thickness $h = 2$. The orientation angles of the bottom and top layers are 30° and −60°, respectively, with respect to the z-direction. The crack has the inclination angle 45° relative to the interface and extends across the entire thickness of the 0° layer. It is a transverse matrix crack since the crack plane is parallel to the fiber axis of the middle layer. The elastic and thermal properties of the graphite/epoxy composite are those given in Tsai and Hahn (1980):

$$E_1 = 181 \text{ Gpa}, \quad E_2 = E_3 = 10.3 \text{ Gpa}, \quad G_{12} = G_{13} = 7.17 \text{ Gpa}, \; G_{23} = 4.02 \text{ Gpa},$$
$$\nu_{12} = \nu_{13} = \nu_{23} = 0.28, \quad \alpha_1 = 0.02\times10^{-6}/°, \quad \alpha_2 = \alpha_3 = 22.5\times10^{-6}/°,$$

where 1 denotes the fiber direction. The length of the model in the x-direction is six times the layer thickness h. Finite element analysis of the model is obtained using the *PDEase* finite element code for a uniform temperature load $\Delta T = -1°$.

The tri-sector wedge at the lower end of the oblique crack is examined. The wedge contais a 180° sector of the 30° ply interfacing with two disjoint sectors of the 0° middle layer. Again, the finite element solution must be purged of its thermal strain, by superposing appropriate complementary solutions, before the finite element results can be used as the collocation data. Once again, the first complementary solution represents free thermal expansions of the separate layers (except for the reaction stress σ_z to the imposed condition $\varepsilon_z = 0$). This introduces incompatibility of the tangential strains ε_x and γ_{xz} between the differently oriented layers. A second complementary solution is required to compensate for the discontinuity of the two tangential strain components across the interface of the middle and bottom layers. This can be achieved by applying uniform stresses $\sigma_x = 0.14514$ MPa and $\tau_{xz} = 0.16475$ Mpa in the bottom layer only, while imposing no stress in the cracked middle layer.

The two 0° sectors are transversely isotropic and therefore degenerate. The sectors have triple material eigenvalues $\mu = \pm i$, so that Eq. (3.3) must be used in these sectors. The material eigenvalues of the 30° sector are distinct and purely imaginary, i.e., $\pm 0.959036i$, $\pm 1.093396i$ and $\pm 2.596064i$.

Eigenvalues of the wedge with real parts smaller than 5 are used to form a truncated eigenseries with 27 terms. These eigenvalues all lie very close to the real axis of the λ-plane. Each positive integer is a *triple* root with three independent eigensolutions. The noninteger eigenvalues appear to cluster together in groups of three near the points $n - 0.5$ on the real axis ($n = 1, 2, 3, \ldots$). Those of the first group are all real: 0.489119, 0.505679 and 0.613022. Notice that the first one is smaller than 0.5, so that the corresponding eigensolution has a stronger singularity than that of an interface crack. The second group consists of one real and two complex conjugate roots: 1.457716 and $1.528059 \pm 0.0266804i$. The third group has real eigenvalues only: 2.448084, 2.465270, 2.586912.

Results of the interfacial stresses on the right and left interface ($\theta = 0°$ and $\theta = -180°$) are shown, respectively, in Figs. 5-7 and Figs. 8-10. Again, the elasticity solution is normalized with respect to the power of the asymptotic solution and the distance r from the singularity is shown in the logarithmic scale. For all three componenents of the interfacial stresses, the differences between the asymptotic and the elasticity solutions remain significant even as r decreases to the order 10^{-10}. For the shearing stresses τ_{xy} and τ_{yz}, the differences become negligible only as $r < 10^{-15}$. For the normal stress σ_y on the left interface, the approach of the elasticity solution to the asymptotic solution is extremely slow (Fig. 8).

Notice that in the region $r < 10^{-3}$, the interlaminar stresses given by the sum of the three leading terms of the eigenseries (shown by dashed curves in the figures) are quite close to the full sum of 27 terms. These leading terms correspond to the first group of three eigenvalues close to 0.5. This does not mean that the higher eigensolutions may be neglected in the presenet analyisis. In fact, the coefficients of the eigensolutions associated with the three groups of non-integer eigenvalues clustered around 0.5, 1.5 and 2.5 are: 16.6138, 0.2228, −12.9303, 4.7875, $-0.5314+2.4302i$, $-0.5314-2.4302i$, −7.4868, 8.5630 and −2.6437. The coefficients of the eigensolutions of the third group (with eigenvalues close to 2.5) are quite significant in magnitude. A truncated eigenseries which does not contain these three eigensolutions will significantly change the results of collocation. It is found that the eigenseries consisiting only of twelve eigensolutions with $\mathrm{Re}[\lambda] \le 2$ gives reasonably accurate results for the interlaminar shearing stresses on the two interfaces, but causes significant errors in the peeling stress across the left interface.

REFERENCES

A. Y. Kuo (1989), "Thermal stresses at the edge of a bimetallic thermostat," *J. Appl. Mech.*, Vol. 56, pp. 585-589.

T. C. T. Ting (1996), *Anisotropic Elasticity: Theory and Application.* Oxford University Press, New York, NY.

S. W. Tsai and H. T. Hahn (1980), *Introduction to Composite Materials*, Technomic, Westport, CT.

A. S. D. Wang and F. W. Crossman (1977a), "Some new results on edge effect in symmetric composite laminates," *J. Composite Materials*, Vol 11, pp. 92-106.

A. S. D. Wang and F. W. Crossman (1977b), "Edge effects on thermally induced stresses in composite laminates," *J. Composite Materials*, Vol 11, pp. 300-312.

S. S. Wang and I. Choi (1982), "Boundary-layer effects in composite laminates: Part 1-- Free-edge stress singularities," *J. Appl. Mech.* Vol. 49, pp. 541-548.

S. S. Wang and I. Choi (1982), "Boundary-layer effects in composite laminates: Part 2-- Free-edge stress solutions and basic characteristics," *J. Appl. Mech.*, Vol. 49, pp. 549-560.

S. S. Wang and F. G. Yuan (1983), "A hybrid finite element approach to laminate elasticity problems with stress singularities," *J. Appl. Mech.*, Vol. 50, pp. 835-844.

W.-L. Yin (1996), "Singular solutions of interfacial stresses in multilayered structures," in *Structural Analysis in Microelectronics and Fiber Optics*, ASME EEP-Vol. 16, E. Suhir Editor, ASME, New York, pp. 59-65.

W.-L. Yin (1997), "A general analysis methodology for singularities in composite structures," *Proc. AIAA/ASME/ASCE/AHS/ASC 38th SDM Conference*, April 7-10, 1997, Kissimere, FL, pp. 2238-2246.

W.-L. Yin, K.C. Jane and C.-C. Lin (1997), "Singular solutions of multi-material wedges under thermomechanical loading," *Analysis and Design Issues for Modern Aerospace Vehicles -- 1997*, ASME AD-Vol. 55, G. J. Simitses Ed., ASME, New York, pp. 159-166.

W.-L. Yin (1998), "Delamination: Laminate analysis and fracture mechanics," *Fatigue and Fracture of Engng. Materials & Struc.*, Vol. 21, pp. 509-520.

W.-L. Yin (1999a), "Deconstructing plane anisotropic elasticity, Part I: The latent structure of Lekhnitskii's formalism," to appear in *Intl.J. Solids Struc.*

W.-L. Yin (1999b), "Deconstructing plane anisotropic elasticity, Part II: Stroh's formalism sans frills," to appear in *Intl.J. Solids Struc.*

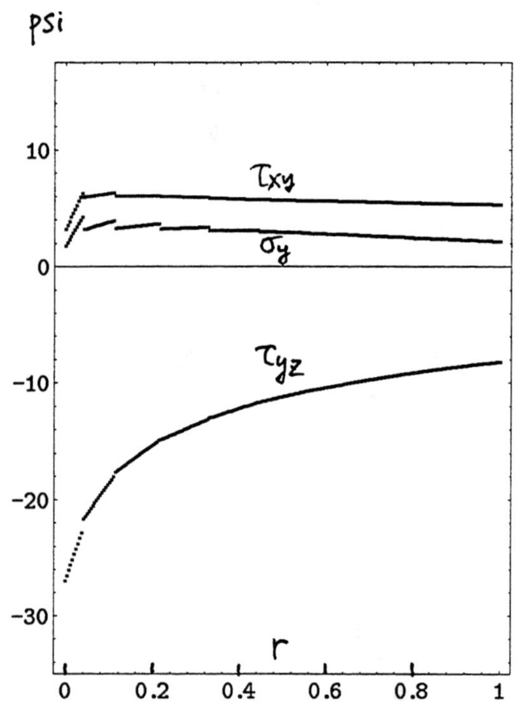

Fig. 1

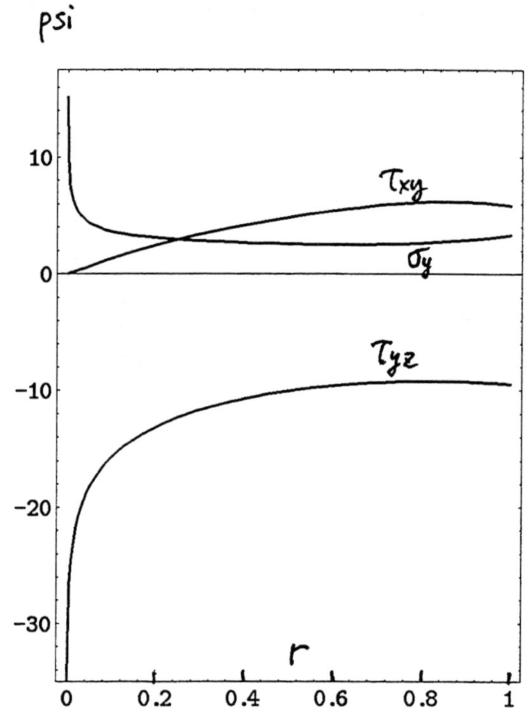

Fig. 2

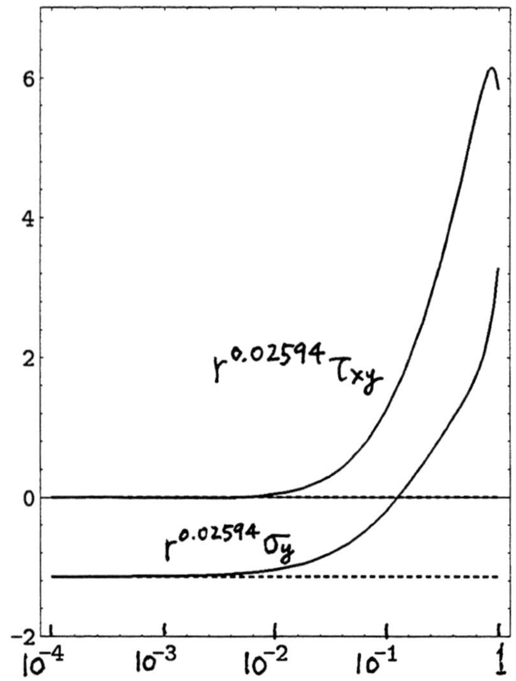

Fig. 3

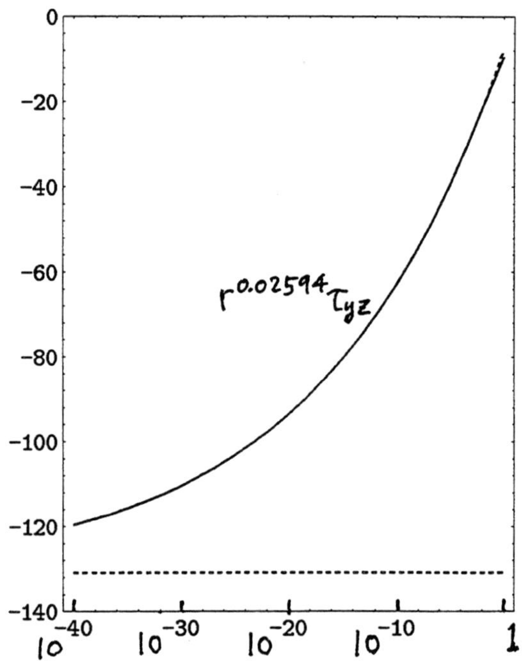

Fig. 4

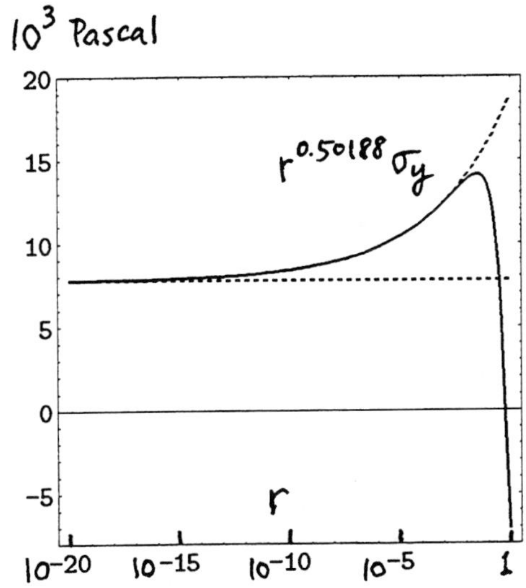

Fig. 5

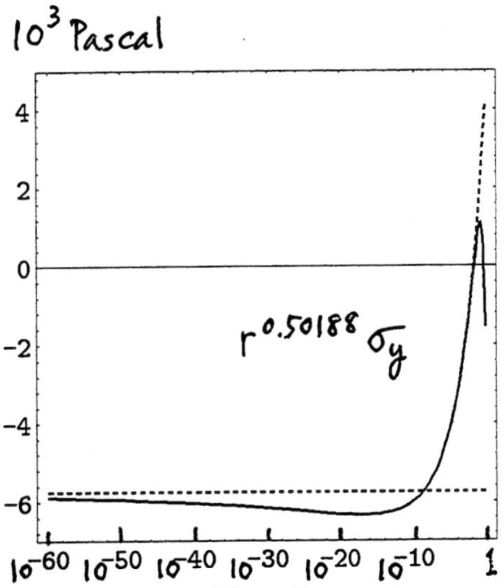

Fig. 8

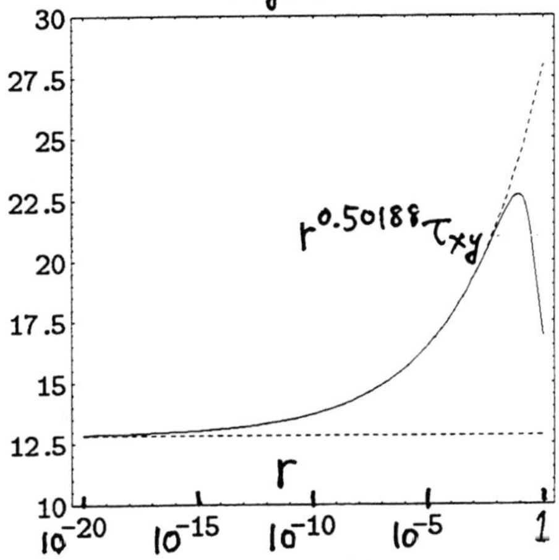

Fig. 6

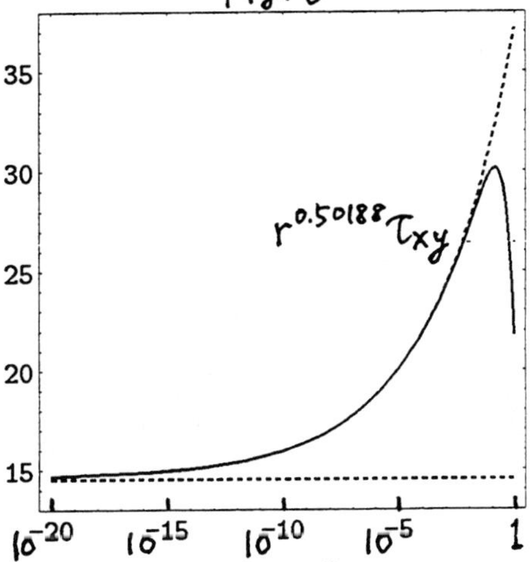

Fig. 9

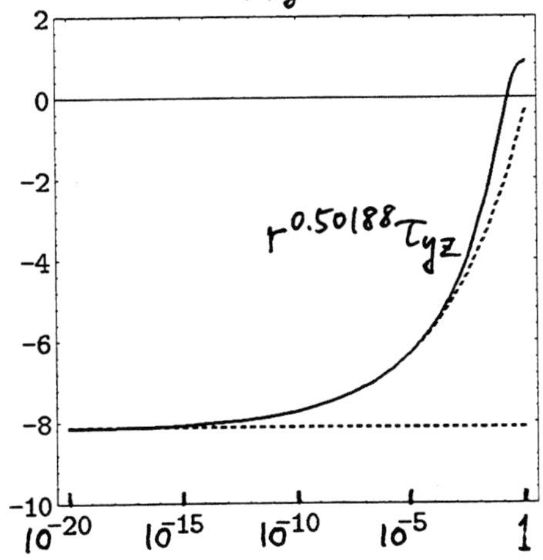

Fig. 7

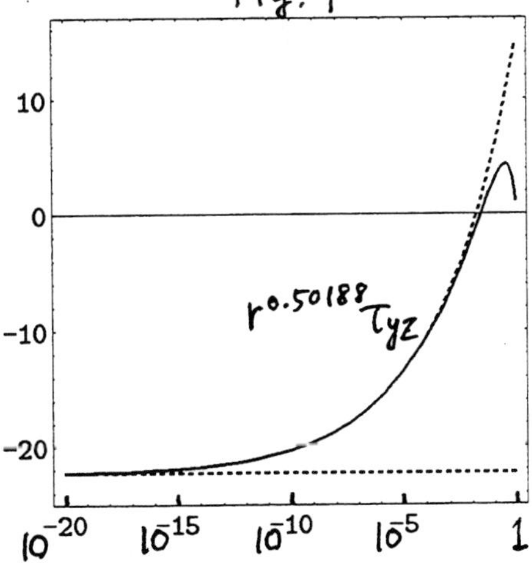

Fig. 10

AMD-Vol. 235, Thick Composites for Load Bearing Structures
ASME 1999

STIFFNESS AND STRENGTH PROPERTIES FOR BASIC SANDWICH MATERIAL CORE TYPES

Beomkeun Kim*
Stanford University
Stanford, California, 94305 United States

Richard M. Christensen
Lawrence Livermore National Laboratory, P. O. Box 808
Livermore, California 94551, United States
Phone: (925) 422-7136, Fax: (925) 423-2405
christensen6@llnl.gov

ABSTRACT

Three basic core material types for sandwich structure applications are studied. The three two-dimensional pattern types are: honeycomb, triangular cells, and a new configuration involving star type cells. The specific critical properties of stiffness and strength type are identified and studied, both theoretically and experimentally.

INTRODUCTION

Sandwich structures have been used in numerous aspects of aircraft construction as well as in many other applications. Typical sandwich structures are made up of two thin, stiff, strong faces separated by a very lightweight material known as the core. By choosing appropriate core materials based on intended applications, not only are sandwich panels able to achieve high stiffness, and strengths comparable to those of single solid panels, but also great savings in weight.

Many types of low density materials that satisfy the requirements of sandwich structures are used as core materials, and one of the most popular cores is the hexagonal core. This core has a two-dimensional hexagonal pattern made of thin walled structures in one plane. Because cores with two-dimensional patterns such as hexagonal core provide a direct connection between the face panels, they generally show good structural efficiency.

When a material has hexagonal symmetry like a hexagonal core (Figure 1), the stiffness matrix relating stresses and strains has five independent constants. This reveals that the hexagonal core can be considered transversely isotropic and has a plane of isotropy. Gibson and Ashby [1997] have extensively studied the honeycomb form.

A core with a triangular pattern is also isotropic in the plane. One triangle shows only triangular symmetry. However, if six triangles are combined, this pattern has hexagonal symmetry as shown in Figure 1. in other words, this triangular core is isotropic in this plane. Christensen [1995] suggested a new two-dimensional pattern which has six-pointed star-shaped cells adjacent to hexagonal cells. This pattern also has hexagonal symmetry and is isotropic in the plane (Figure 1).

In the current study, the primary objective is to examine and compare the relevant mechanical characteristics of behavior for the three core types having isotropic patterns—the triangular core, the hexagonal core and the starcell core—as related to applications in sandwich structures.

Results

The mechanical characteristics of three types of cores with two-dimensional isotropic patterns—triangular, hexagonal, and starcell—were studied as related to applications in sandwich structures. The Young's modulus, shear modulus, and Poisson's ratio were calculated for the three core types in the direction normal to the faces. The compressive buckling strength and shear buckling strength were calculated for the three core types by modeling each cell wall of the core as a plate under compressive or shear load. To verify this model, tests were conducted on scaled specimens to measure the compressive buckling strength of each core. The bending flexibility's of the three cores were also studied. Compliances for the three cores were measured using flexural tests. Tests were performed on each core type in which the deflection of a circular core sample loaded at its center was measured. The three isotropic core patterns exhibited distinct characteristics, as shown in Figure 2. The results are expressed as a function of the relative density of the core material, and discussed in more detail by Kim and Christensen [1999].

REFERENCES

Gibson, L. J. and Ashby, M., Cellular Solids Structure and Properties, (Cambridge University Press, Cambridge, 1997).

Christensen, R. M., "The hierarchy of microstructures for low density materials," ZAMP 46 (1995) 5506-5521.

Kim, B. and Christensen, R. M., "Basic Two-Dimensional Core Types for Sandwich Structures," Int. J. Mechanical Sciences, in press.

- Objective

  Detailed study of three prototype 2-D core forms for sandwich structures

- Basic 2-D Core Forms

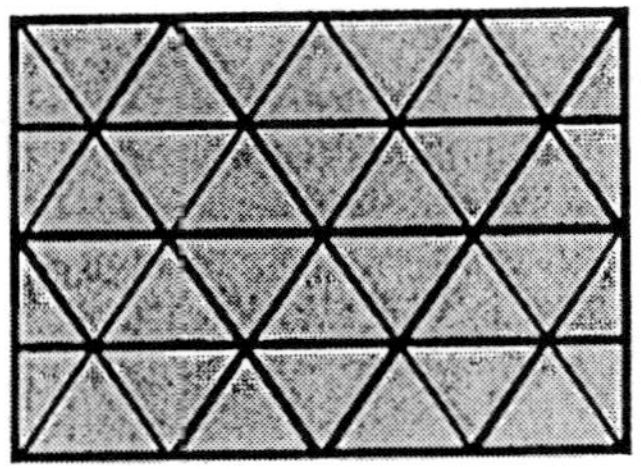

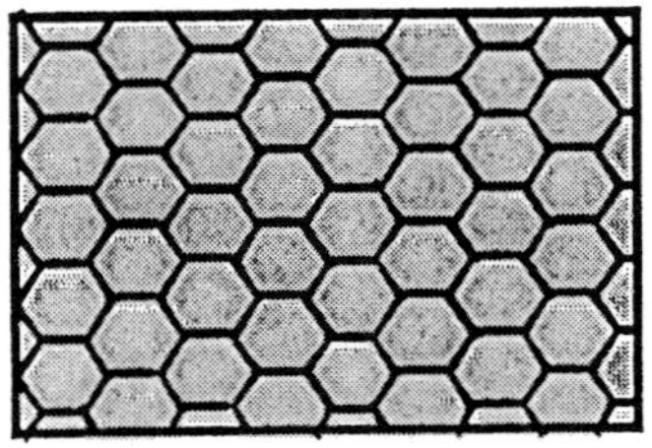

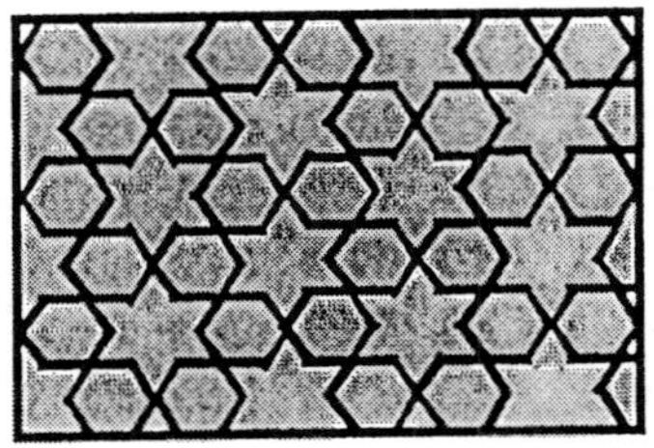

- Approach

  Theoretical and Experimental investigation of three dimensional stiffness and strength properties

- Critical Properties

  In-plane moduli, shear G and bulk K

  Out of plane shear strength

Figure 1 Patterns

Stiffness Theory

	Triangular	Hexagonal	Star
G/E_m	$\frac{1}{8}\left(\frac{\rho}{\rho_m}\right)$	$\frac{3}{8}\left(\frac{\rho}{\rho_m}\right)^3$	$\frac{3}{16}\left(\frac{\rho}{\rho_m}\right)^3$
K/E_m	$\frac{1}{4}\left(\frac{\rho}{\rho_m}\right)$	$\frac{1}{4}\left(\frac{\rho}{\rho_m}\right)$	$\frac{9}{16}\left(\frac{\rho}{\rho_m}\right)^3$

* ρ - density

Compressive Strength, Shear Strength Theory

	Triangular	Hexagonal & Star
σ_{cr}	$\frac{1}{9}\beta_n\frac{E_m}{1-\nu_m^2}\left(\frac{\rho}{\rho_m}\right)^3$	$\beta_n\frac{E_m}{1-\nu_m^2}\left(\frac{\rho}{\rho_m}\right)^3$
τ_{cr}	$\frac{1}{9}\beta_s\frac{E_m}{1-\nu_m^2}\left(\frac{\rho}{\rho_m}\right)^3$	$\beta_s\frac{E_m}{1-\nu_m^2}\left(\frac{\rho}{\rho_m}\right)^3$

* β_n, β_s from boundary conditions

Stiffness Experiment

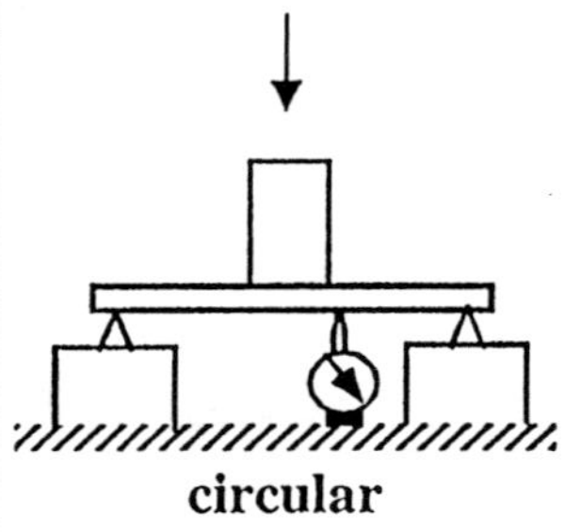

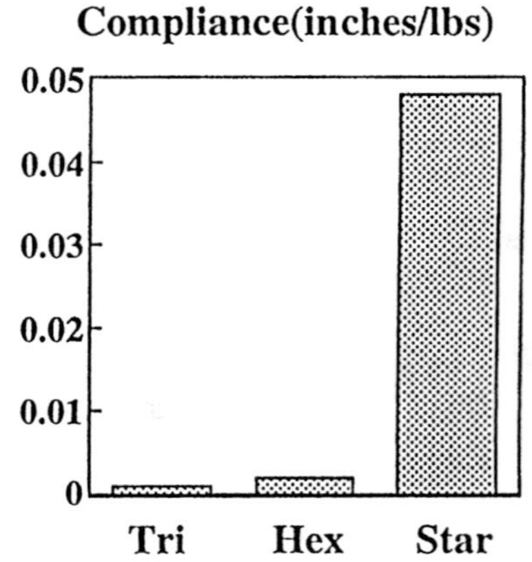

Strength Experiment

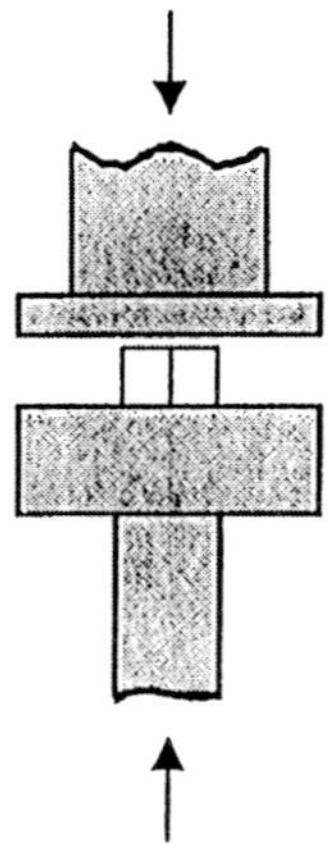

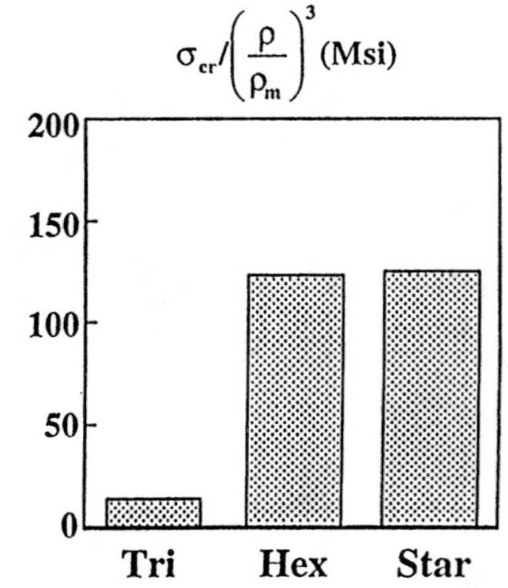

Figure 2 Results

AMD-Vol. 235, Thick Composites for Load Bearing Structures
ASME 1999

POSTBUCKLING AND GROWTH BEHAVIOR OF FACE-SHEET DELAMINATIONS IN SANDWICH COMPOSITES

George A. Kadomateas

School of Aerospace Engineering, Georgia Institute of Technology
Atlanta, Georgia 30332-0150, USA

ABSTRACT

A bending load on a sandwich beam consisting of two fiber-reinforced sheets (skins) separated by a low stiffness core is equivalent to a compressive force on one face and a tensile force on the other. Should an interface crack between the layers of the composite face-sheet or between the core and the composite face-sheet exists on the compressively loaded face, local buckling and subsequently growth of this interface crack (delamination) may occur. This may also naturally occur under pure compression. In this study, the buckling, postbuckling and delamination growth behavior is studied through a procedure that is based on the large deflections of the delaminated layer. First, the solution is derived for the postbuckled states by using the elastica theory to model the large deflections of the thin delaminated layer and the sandwich beam theory (for unsymmetric sections) that includes transverse shear, to model the (relatively modest) deflections of the rest of the structure. This postbuckling solution is subsequently used to study the growth of the mixed mode interface crack through an energy release rate approach. The consequences of the elastic mismatch between core and face sheet are discussed and illustrative results are presented for several sandwich construction configurations.

INTRODUCTION

Delaminations (layer interface cracks) constitute a common failure phenomenon in laminated composites and they are most easily introduced from impact loads. These delaminations may deteriorate the performance of the structure under compressive loading (e.g., Yin et al, 1986; Simitses et al, 1985). A large number of studies on the behavior of delamination buckling and postbuckling in composites have been carried out by many researchers, e.g. Chai et al (1981) by using a one dimensional model, Kardomateas (1990) by conducting monotonic compressive tests, Kardomateas (1993) by using elastica theory to account for large deformations during postbuckling, Kardomateas et al (1995) by studying both experimentally and analytically the fatigue growth of delaminations during cyclic compression, etc.

Although the general principles are not very different, delamination failure in sandwich structures is just beginning to be explored in detail. In this regard, differences in the behavior of delamination buckling and postbuckling within a sandwich structure from that of a laminated composite structure arise due to the fact that the substrate in a delaminated sandwich structure includes a much different kind of material, namely a transversely flexible core made of foam or low strength honeycomb. To this extent, the contribution of the shear stresses and shear deformations of the core are expected to be noteworthy and therefore should be included in the formulation.

A typical sandwich structure is composed of two thin composite laminated faces and a thick soft core made of foam or low strength honeycomb. Due to its exceptional properties, mainly high stiffness and strength with little resultant weight penalty, sandwich structures have been used in aircraft, marine and other types of structures. Research into sandwich structural behavior and failure modes can be traced following World War II in a rather sporadic fashion but intensified in the 1990s esp. with regard to proper modelling of the core through high order theories (e.g., Kant and Patil, 1991; Hunt and Da Silva, 1990a,b; Frostig et al., 1992 and Frostig and Baruch, 1993).

Although these high order theories are expected to render most accurate results, they involve considerable effort in addressing the complexities of the formulation of the problem of postbuckling of delaminated beams, and therefore, in the present work, a simple beam theory including transverse shear, peroperly formulated for an unsymmetric sandwich section (meaning face sheets not of the same geometry and/or material) is used to model the substrate and base parts which undergo modest bending deformations, whereas the nonlinear elastica theory is used to model the delaminated layer, which undergoes large deflections during postbuckling. The same approach can be used to study either a delamination within the face sheet or a debond at the interface between the face sheet and the core.

Fracture mechanics principles are subsequently used to study the possible growth of the delamination during postbuckling. In particular, changes of the total strain energy in a sandwich panel between successive states would lead to the calculation of the energy release rate, which can be used to predict delamination growth. Therefore, this work would also focus on investigating the growth characteristics.

FORMULATION

The postbuckled shape is shown in Fig. 1. A sandwich beam with layered composite top and bottom face-sheets of thickness f_1 and f_2, respectively, and a core of thickness c, has a delamination of thickness h, symmetrically located in the top face-sheet. Over the region of the delamination, of length ℓ, the sandwich beam consists of two parts: the delaminated layer of the upper face sheet (referred to as the "delaminated part", of thickness h) and the part below the delamination ("substrate part", of thickness $f_1 - h + c + f_2$, which includes the core and the lower face sheet). The region outside the deamination is referred to as the "base part" and consists of the entire section of the sandwich beam, i.e., of thickness $f_1 + c + f_2$.

To describe the deformation of the delaminated part, which undergoes large deflections and therefore geometric nonlinearities should be taken into account, we use the exact theory of plane deformation of members that are restrained elastically at the ends by means of concentrated forces and moments (elastica theory). Due to the small thickness of the delaminated layer, the effect of transverse shear forces on the deformation of the delaminated layer is not included. Also, the elastica formulas are applied with a bending stiffeness, $(EI)_d$, corresponding to the modulus in the $x \equiv 1$ axis, i.e., $(EI)_d = E_{f1}BH^3/[12(1-\nu_{13}\nu_{31}]$, where E is used to denote the extensional modulus and ν to denote the Poisson's ratio and B is the width; also $z \equiv 3$. For the rest of the sandwich beam, i.e. the substrate and the base parts, since the flexural deflections are small, we use the linear theory of bending of an unsymmetric sandwich beam (meaning face sheets not of the same geometry and/or material). This theory is used with transverse shear included due to the existence of the low modulus, relatively thick, core. The conditions of geometric continuity that will be enforced play an important role in the realization of the equilibrium states.

Delaminated Part

Let us first focus on the delaminated layer. In the buckled form, which is assumed to be symmetrical, denote by P_d and M_d the end forces and moments. In the following equations, the quantities at the right end are used (see Fig. 1). The moments and angles are assumed to be positive clockwise. Two variables play an important role: the end-amplitude for the compressed film, Φ_d, and the distortion parameter, β, which is the angle of tangent rotation at the inflection point. These are the generalized coordinates of deformation. In terms of these two quantities, relations between the end stress resultants P_d, M_d, as well as other coordinates of deformation such as the end slope, θ and total flexural contraction, e_d, can be found. The relations require the use of elliptic integrals. Denote:

$$k = \sin(\beta/2) \ . \tag{1a}$$

Then, the first elliptic integral, $F(\Phi, k)$, is defined by:

$$F(\Phi, k) = \int_0^{\Phi} \frac{d\phi}{\sqrt{1 - k^2 \sin^2 \phi}} \ , \tag{1b}$$

and the second elliptic integral, $E(\Phi, k)$, by:

$$E(\Phi, k) = \int_0^{\Phi} \sqrt{1 - k^2 \sin^2 \phi} \ d\phi \ . \tag{1c}$$

The values of these integrals at the end, $F(\Phi_d, k)$, $E(\Phi_d, k)$, are denoted by F_d and E_d, respectively. The following five relations define the characteristics of the postbuckled delaminated layer (Britvek, 1973):

Axial force:

$$P_d = 4(EI)_d F_d^2/\ell^2 \ . \tag{2a}$$

End moments:

$$M_d = 4\frac{(EI)_d}{\ell} k F_d \cos \Phi_d \ . \tag{2b}$$

End rotations:

$$\theta = 2\arcsin(k \sin \Phi_d) \ . \tag{2c}$$

Flexural contraction:

$$e_d = 2\ell\left[1 - (E_d/F_d)\right] \ . \tag{2d}$$

Flexural elastic energy:

$$U_d = 8\frac{(EI)_d}{\ell} F_d \left[E_d - F_d \cos^2(\beta/2)\right] \ . \tag{2e}$$

Furthermore, the deflection at the middle, normal to the central line of thrust (direction of axial forces P_d), denoted by Y_{dm}, is found from:

$$Y_{dm} = 2k\sqrt{\frac{(EI)_d}{P_d}}\,(1 - \cos \Phi_d) \ . \tag{2f}$$

It should be noted that at each point x there corresponds a value of the variable Φ (amplitude for compressive members); the value at $x = \ell/2$ is Φ_d. Furthermore, at the inflection point, where the slope $\theta = \beta$, the value is $\Phi = \pi/2$ and at the middle $x = 0$, where $\theta = 0$, $\Phi = 0$. With these remarks in mind, the deflections at all other points are given from:

$$y_d(x) = 2k\sqrt{\frac{(EI)_d}{P_d}}\left[\cos \Phi(x) - \cos \Phi_d\right] \ , \tag{2g}$$

where $\Phi(x)$ is found from the implicit relation:

$$F\left[\Phi(x), k\right] = 2F_d x/\ell \ . \tag{2e}$$

<u>*Substrate and Base Parts*</u>

For the substrate and base parts, since the induced bending is moderate, we use the linear theory of bending of sandwich beams under applied end moments and axial forces, which includes transverse shear. These parts consist of a soft core of thickness c, for example, a hardened polymeric foam or honeycomb, and two stiff faces (skins). The contribution of the longitudinal normal stresses in the core is small compared with those in the skins. Consequently, the shear stress is nearly uniform through the thickness of the core. Since the skins are thin, the shear stresses in the core carry most of the shear force, and the shear deformation of the core is very important. The shear deformations can be taken into account by relaxing the assumption that the plane cross sections remain normal to the deflected beam axis, that is, the slope θ of the deflected beam axis is no longer required to be equal to the rotation ψ of the cross section, the difference being the equivalent for the section shear angle, γ_{eq} (Timoshenko beam theory assumption, Timoshenko and Gere, 1961):

$$\gamma_{eq} = \theta - \psi = \frac{\alpha V}{G_c A_c} \ , \tag{3}$$

in which V represents the shear force that is normal to the deflected beam axis and rotates as the beam deflects, and G_c and A_c are the shear modulus and the cross sectional area of the core. Furthermore, α is the shear correction coefficient, which is calculated in the Appendix for an arbitrary unsymmetric sandwich structure from strain energy considerations and takes into account the nonuniform distribution of the shear stresses throughout the cross section and the contribution of the skins. The equivalent rigidity of the entire section, $(EI)_{eq}$, is also given in the Appendix for an arbitrary asymmetric sandwich construction.

At this point it should be noted that the substrate section is a sandwich beam with a top skin $f_{1s} = f_1 - h$, the core and bottom skins being the same as in the parent beam. The base part is obviously a sandwich section identical to the parent beam (Fig. 1).

Using $V = -M'$ and $\theta = w'$ and $\psi' = M/(EI)_{eq}$, where M is the bending moment, gives the total curvature w'' as the sum of the flexural curvature and the curvature due to shear:

$$w'' = \frac{M}{(EI)_{eq}} + \left(\frac{-\alpha M'}{G_c A_c}\right)' \ . \tag{4a}$$

Now since $M(x) = M_0 - Pw$ (Fig. 2), where M_0 is the end moment (M_s or M_b for the substrate or base parts, respectively), we obtain after substitution in (4a):

$$w''\left[1 - \frac{\alpha P}{(G_c A_c)}\right] + \frac{P}{(EI)_{eq}} w = \frac{M_0}{(EI)_{eq}} \ . \tag{4b}$$

Then, the rotation of the cross section is:

$$\psi = w' + \frac{\alpha M'}{G_c A_c} = \left(1 - \frac{\alpha P}{G_c A_c}\right) w' \ , \tag{4c}$$

and the bending moment is:

$$M = (EI)_{eq} \left(1 - \frac{\alpha P}{G_c A_c}\right) w'' \ . \tag{4d}$$

Definining

$$k^2 = \frac{P}{(EI)_{eq}[1 - \alpha P/(G_c A_c)]} \ , \tag{4e}$$

gives the differential equation for the deflections of the substrate and the base parts in the form:

$$w'' + k^2 w = \frac{M_0}{(EI)_{eq}[1 - \alpha P/(G_c A_c)]} \tag{4f}$$

Substrate Part

For the substrate part, the deflections that satisfy $w = 0$ at $x = 0, \ell$, and the end slope $w'(0) = -w'(\ell) = \theta$ (Fig. 2), are:

$$w(x) = \frac{\theta}{k_s \sin(k_s \ell/2)} \left[\cos\left(\frac{k_s \ell}{2} - k_s x\right) - \cos\frac{k_s \ell}{2}\right] \ , \tag{5a}$$

where k_s is found by using (4e) with P_s, $(EI)_s$ and α_s for the sandwich beam which constitutes the substrate part, by substituting into (A2) and (A6) of the Appendix, resp., i.e.

$$k_s^2 = \frac{P_s}{(EI)_s[1 - \alpha_s P_s/(G_c A_c)]} \ . \tag{5b}$$

The end moment, M_s, is found by using (4d):

$$M_s = -(EI)_s \left(1 - \frac{\alpha_s P_s}{G_c A_c}\right) \theta k_s \cot(k_s \ell/2) \ . \tag{5c}$$

The corresponding shortening is:

$$e_s = \frac{1}{2} \int_0^\ell w_s'^2 dx = \frac{\theta^2 (\ell k_s - \sin k_s \ell)}{4 k_s \sin^2(k_s \ell/2)} \ , \tag{5d}$$

and the energy due to bending and transverse shear [see Appendix, eqn (A4)], is (since $V = -M'$):

$$U_s = \frac{1}{2(EI)_s} \int M^2 dx + \frac{\alpha_s}{2 G_c A_c} \int M'^2 dx \ ,$$

or,

$$U_s = (EI)_s \left(1 - \frac{\alpha_s P_s}{G_c A_c}\right)^2 \frac{\theta^2 k_s}{4 \sin^2 k_s \ell/2} \left[(k_s \ell + \sin k_s \ell) + \frac{\alpha_s (EI)_s}{G_c A_c} k_s^2 (k_s \ell - \sin k_s \ell)\right] \ . \tag{5e}$$

Base Part

For the base part, subject to a slope $-\theta$ at $x = 0$, and having a clamped end at $x = b$ (Fig. 2), i.e., $w_b'(0) = -\theta$ and $w_b'(b) = 0$, the deflections that satisfy (4f) are:

$$w_b(x) = -\frac{\theta}{k_b \sin k_b b} [\cos k_b (b - x) - \cos k_b b] \ , \tag{6a}$$

where k_b is

$$k_b^2 = \frac{P}{(EI)_b[1 - \alpha_b P/G_c A_c]} \ . \tag{6b}$$

The end moment, M_b, is:

$$M_b = (EI)_b \left(1 - \frac{\alpha_b P}{G_c A_c}\right) w'' \mid_{x=0} = (EI)_b \left(1 - \frac{\alpha_b P}{G_c A_c}\right) \theta k_b \cot(k_b b) \ . \tag{6c}$$

The flexural shortening for both the left and right base parts is:

$$e_b = 2 \int_0^b \frac{w_b'^2}{2} dx = \frac{\theta^2 (2k_b b - \sin 2k_b b)}{4k_b \sin^2 k_b b} \ , \tag{6d}$$

and the energy due to bending and transverse shear is:

$$U_b = \frac{1}{2(EI)_b} \int M^2 dx + \frac{\alpha_b}{2G_c A_c} \int M'^2 dx \ ,$$

or, by use of (4d) and (6a),

$$U_b = \frac{(EI)_b}{2} \left(1 - \frac{\alpha_b P}{G_c A_c}\right)^2 \frac{\theta^2 k_b}{4 \sin^2 k_b b} \left[(2k_b b + \sin 2k_b b) + \frac{\alpha_b k_b^2 (EI)_b}{G_c A_c} (2k_b b - \sin 2k_b b) \right] \ . \tag{6e}$$

Equilibrium at the Common Section

The following conditions of force and moment equilibrium at the right common section should be fulfilled (Fig. 1):

$$P = P_d + P_s \ , \tag{7a}$$

$$M_d + M_s + P_d (f_1 - h + c + f_2)/2 - P_s h/2 = M_b \ . \tag{7b}$$

Moreover, a condition of compatible shortening (which includes shortening due to bending and that due to compression) for the delaminated and substrate parts reads:

$$e_d + \left(1 - \nu_{13}^{(d)} \nu_{31}^{(d)}\right) \frac{P_d \ell}{E_{f1} h B} - e_s -$$

$$- \frac{\left(1 - \nu_{13}^{(s)} \nu_{31}^{(s)}\right) P_s \ell}{[E_{f1}(f_1 - h) + E_c c + E_{f2} f_2] B} = (f_1 + c + f_2) \mid \theta \mid \ . \tag{7c}$$

In fact, the total shortening of the beam, δ, which is useful because it can easily be obtained as experimental data, is expressed by:

$$\delta = e_b + 2 \frac{\left(1 - \nu_{13}^{(b)} \nu_{31}^{(b)}\right) Pb}{(E_{f1} f1 + E_c c + E_{f2} f_2) B} + e_d +$$

$$+ \left(1 - \nu_{13}^{(d)} \nu_{31}^{(d)}\right) \frac{P_d \ell}{E_{f1} h B} - \mid \theta \mid (f_1 - h + c + f_2) \ . \tag{8}$$

The total strain energy of the system is that due to both bending and compression of the different parts and is expressed as:

$$U_{tot} = U_d + \frac{P_d^2 \ell}{2E_{f1} h B} + U_s + \frac{P_s^2 \ell}{2[E_{f1}(f_1 - h) + E_c c + E_{f2} f_2] B} +$$

$$+ U_b + \frac{P^2 b}{(E_{f1} f1 + E_c c + E_{f2} f_2) B} \tag{9}$$

where U_d, U_s and U_b are given by eqs (2e), (5e) and (6e).

Solution Procedure

For a given end rotation angle, θ, we obtain the following system of two nonlinear algebraic equations in β (the tangent rotation at the inflection point of the delaminated film) and P_s (axial load for the substrate part) by substituting eqs (2), (5) and (6) into (7):

From (7b):

$$-4\frac{(EI)_d}{\ell}F_d\cos\Phi_d\sin(\beta/2)-(EI)_s\left(1-\frac{\alpha_sP_s}{G_cA_c}\right)\theta k_s\cot(k_s\ell/2)-$$

$$-(EI)_b\left(1-\frac{\alpha_bP}{G_cA_c}\right)\theta k_b\cot(k_bb)+2\frac{(EI)_d}{\ell^2}(f_1-h+c+f_2)F_d^2-P_s\frac{h}{2}=0\,, \qquad (10a)$$

and from (7c):

$$2\ell\left(1-\frac{E_d}{F_d}\right)-\frac{\theta^2\left(k_s\ell-\sin k_s\ell\right)}{4k_s\sin^2(k_s\ell/2)}+4\left(1-\nu_{13}^{(d)}\nu_{31}^{(d)}\right)\frac{(EI)_dF_d^2}{E_{f1}\ell hB}-$$

$$-\frac{\left(1-\nu_{13}^{(s)}\nu_{31}^{(s)}\right)P_s\ell}{[E_{f1}(f_1-h)+E_cc+E_{f2}f_2]B}-(f_1+c+f_2)\mid\theta\mid=0\,, \qquad (10b)$$

where, in these equations, k_s, P and k_b are determined directly for a given P_s as follows: k_s is found from (5b), and from (7a),

$$P=P_s+4\frac{(EI)_d}{\ell^2}F_d^2\,, \qquad (10c)$$

and then k_b is found from (6b).

Also, for a given β, in these equations, from (2c) and (1a), Φ_d is given in terms of θ and β by:

$$\Phi_d=\arcsin\left[\frac{\sin(\theta/2)}{\sin(\beta/2)}\right]\,, \qquad (10d)$$

and F_d, E_d are defined from the elliptic integrals:

$$F_d=\int_0^{\Phi_d}\frac{d\phi}{\sqrt{1-\sin^2(\beta/2)\sin^2\phi}}\,, \qquad (10e)$$

$$E_d=\int_0^{\Phi_d}\sqrt{1-\sin^2(\beta/2)\sin^2\phi}\,d\phi\,. \qquad (10f)$$

Thus, for a given end rotation angle, θ, the equilibrium state is found by numerically solving eqs (10a) and (10b) for β and P_s. Notice that the angle $\theta<0$ in Fig. 2. Moreover, in solving the system, the angle β is varying between θ and $\pi/2$, and the load P_s between zero and the critical load of the substrate. Notice that the end amplitude Φ_d is $\pi\leq\Phi_d\leq 3\pi/2$. Thus, the equilibrium states are found by determining the values of β and P_s that satisfy the compatibility condition, eqn (7c), and the equilibrium condition, eqn (7b), at the section where the delamination ends. To solve the system of the two nonlinear equations, we use the Newton-Raphson method, generalized to two dimensions (Press et al, 1989); the method requires the derivatives of the equations with respect to β and P_s, which can be very easily obtained. Results from this formulation are currently being obtained.

Delamination Growth

Let us now consider delamination growth. It is assumed that whether further delamination occurs depends on the magnitude of the fracture energy, defined as the energy required to produce a unit if new delamination. To this extent, a Griffith-type fracture criterion is employed, and the postbuckling solution that has been obtained above can be used to calculate the energy-release rate as a function of the applied axial displacement. This quantity is the differential of the total potential energy with respect to the delamination length.

$$G=-\frac{d\Pi}{Bd\ell}\,. \qquad (11)$$

where B is the width of the sandwich panel, and Π is total potential defined as $\Pi=-U_{tot}/2$.

The above formulas show that the total potential is a function of delamination length ℓ, i.e. $\Pi=\Pi(\ell)$. Thus, at any configuration characterized by a delamination length ℓ, the strain energy and total potential can be calculated for an increment of crack length, $\Delta\ell$, and the crack driving force or energy release rate can be obtained from (11).

ACKNOWLEDGEMENTS

The financial support of the Office of Naval Research, Ship Structures S& T Division, Grant N00014-90-J-1995, and of the Air Force Office of Scientific Research, Grant F49620-98-1-0384, and the interest and encouragement of the Grant Monitors, Dr. Y.D.S. Rajapakse, Dr. Brian Sanders and and Dr. Ozden Ochoa are both gratefully acknowledged.

REFERENCES

Chai H., Babcock C.D. and Knauss W.G. (1981) "One Dimensional Modeling of Failure in Laminated Plates by Delamination Buckling", *Int. J. Solids Structures*, vol. 17, No.11 pp 1069-1083.

Britvek S.J. (1973) *The Stability of Elastic Systems*, Pergamon Press Inc.

Budiansky B. and Rice J.R. (1973) "Conservation Laws and Energy-Release Rates", *Journal of Applied Mechanics* (ASME), vol. 40, pp. 201-203.

Frostig Y. (1992) "Behavior of Delaminated Sandwich Beams with Transversely Flexible Core - High Order Theory", *Composite Structures* vol. 20, pp 1-16.

Frostig Y. and Baruch M. (1993) "Buckling of simply-supported sandwich beams with transversely flexible core - a high order theory", *Journal of ASCE, EM Division*, vol. 119, no. 5, pp. 955-972.

Hunt G.E. and Da Silva L.S. (1990a) "Interaction bending behavior of sandwich beams", *Journal of Applied Mechanics* (ASME) vol. 57, no. 1, pp. 189-196.

Hunt G.E. and Da Silva L.S. (1990b) "Interactive buckling in sandwich structures with core orthotropy", *International Journal of Mechanical Structures and Mechanics* vol. 18, no. 3, pp. 353-372.

Kant T. and Patil H.S. (1991) "Buckling load of sandwich columns with a higher-order theory", *Journal of Reinforced Plastics and Composites* vol. 10, no. 1, pp. 102-109.

Kardomateas G.A. (1993) "The Initial Postbuckling and Growth Behavior of Internal Delaminations in Composite Plates", *Journal of Applied Mechanics* (ASME), vol. 60, pp. 903-910.

Kardomateas G.A. (1990) "Postbuckling Characteristics in Delaminated Kevlar/Epoxy Laminates: An Experimental Study", *Journal of Composites Technology and Research* (ASTM), vol. 12, no. 2, pp. 85-90.

Kardomateas G.A., Pelegri A.A. and Malik B. (1995) "Growth of Internal Delaminations Under Cyclic Compression in Composite Plates", *Journal of the Mechanics and Physics of Solids*, vol. 43, no. 6, pp. 847-868.

Press W.H., Flannery B.P., Teukolsky S.A. and Vetterling W.T. (1989) *Numerical recipes*, Cambridge University Press, Cambridge, MA.

Simitses G.J., Sallam S. and Yin W.L. (1985) "Effect of Delamination on Axially Loaded Homogeneous Laminated Plates", *AIAA Journal*, vol. 23, pp. 1437-1444.

Timoshenko S.P. and Gere J.M. (1961) *Theory of Elastic Stability*, 2nd ed., McGraw-Hill Book Co., New York.

Yin W.L., Sallam S. and Simitses G.J. (1986) "Ultimate axial load capacity of a delaminated beam-plate", *AIAA Journal*, Vol 24, No 1, pp. 123-128.

APPENDIX

An unsymmetric sandwich section is shown in Fig. 3. The section consists of two face sheets of thickness f_1 and f_2 and moduli E_{f1} and E_{f2}, respectively. The core, of thickness c, has an extensional modulus, E_c, and shear modulus G_c.

With respect to the reference axis y through the middle of the core, the neutral axis of the section is defined at a distance e, as:

$$e = \frac{E_{f2}f_2}{E_{f1}f_1 + E_{f2}f_2 + E_c c}\left(\frac{f_2}{2} + \frac{c}{2}\right) - \frac{E_{f1}f_1}{E_{f1}f_1 + E_{f2}f_2 + E_c c}\left(\frac{f_1}{2} + \frac{c}{2}\right), \qquad (A1)$$

and the equivalent flexural rigidity of the sandwich section per unit width, $(EI)_{eq}$, is:

$$(EI)_{eq} = E_{f1}\frac{f_1^3}{12} + E_{f1}f_1\left(\frac{f_1}{2} + \frac{c}{2} + e\right)^2 + E_{f2}\frac{f_2^3}{12} + E_{f2}f_2\left(\frac{f_2}{2} + \frac{c}{2} - e\right)^2 +$$

$$+E_c\frac{c^3}{12}+E_cce^2 \; . \tag{A2}$$

The shear stress can be written in terms of the shear force $V(x)$ in the form:

$$\tau(x,z)=V(x)q(z) \; .$$

Also, due to the sandwich construction, $G=G(z)$, therefore, the shear energy for a beam of width B, is:

$$B\int\int\frac{\tau^2(x,z)}{2G}dxdz=\frac{B}{2}\int V^2(x)\left[\int\frac{q^2(z)}{G(z)}dz\right]dx \; .$$

Now, assume that the shear stresses are carried exclusively by the core in a uniform fashion, then the corresponding equivalent shear stress and strain are

$$\tau_{eq}=\frac{V(x)}{A_c} \; ; \qquad \gamma_{eq}=\frac{\alpha V(x)}{G_cA_c} \; , \tag{A3}$$

where α is a "shear correction coefficient" which takes into account the non-uniform distribution of shear stresses due to the sandwich construction throughout the entire cross section and A_c is the cross sectional area of the core.

Then, the energy due to shear is

$$A_c\int\frac{1}{2}\tau_{eq}\gamma_{eq}dx=\frac{\alpha}{2G_cA_c}\int V^2(x)dx \; , \tag{A4}$$

therefore,

$$\alpha=G_cA_cB\int\frac{q^2(z)}{G(z)}dz \; . \tag{A5}$$

The shear stresses, from simple bending theory, are as follows:
Upper face sheet, $z<0$:

$$\tau_{f1}(z)=\frac{V}{(EI)_{eq}}\frac{E_{f1}}{2}\left[\left(f_1+\frac{c}{2}+e\right)^2-z^2\right] \; .$$

Core, $z<0$:

$$\tau_c(z)=\frac{V}{(EI)_{eq}}\left\{E_{f1}f_1\left(\frac{f_1}{2}+e+\frac{c}{2}\right)+\frac{E_c}{2}\left[\left(\frac{c}{2}+e\right)^2-z^2\right]\right\} \; .$$

Core, $z>0$:

$$\tau_c(z)=\frac{V}{(EI)_{eq}}\left\{E_{f2}f_2\left(\frac{f_2}{2}-e+\frac{c}{2}\right)+\frac{E_c}{2}\left[\left(\frac{c}{2}-e\right)^2-z^2\right]\right\} \; .$$

Lower face sheet, $z>0$:

$$\tau_{f2}(z)=\frac{V}{(EI)_{eq}}\frac{E_{f2}}{2}\left[\left(f_2+\frac{c}{2}-e\right)^2-z^2\right] \; .$$

Therefore, if we define

$$a_i=f_i+\frac{c}{2}+(-1)^{i+1}e \; ; \quad b_i=\frac{c}{2}+(-1)^{i+1}e \; ; \quad c_i=\frac{f_i}{2}+\frac{c}{2}+(-1)^{i+1}e \; ,$$

then the shear correction coefficient is found from (A5) as:

$$\alpha=G_cA_cB\sum_{i=1,2}\frac{E_{fi}^2}{4(EI)_{eq}^2G_{fi}}\left[a_i^4f_i-\frac{2}{3}a_i^2\left(a_i^3-b_i^3\right)+\frac{1}{5}\left(a_i^5-b_i^5\right)\right]+$$

$$+\frac{E_{fi}^2}{(EI)_{eq}^2 G_c}\left[f_i^2 c_i^2 b_i + \frac{2}{15}\frac{E_c^2}{E_{fi}^2} b_i^5 + \frac{2}{3}\frac{E_c}{E_{fi}} f_i c_i b_i^3\right] . \tag{A6}$$

Regarding the Poisson's ratios used in this formulation, a "rule of mixtures" approach is followed, for example,

$$\nu_{13}^{(b)} = \left(\nu_{13}^{(f1)} f_1 + \nu_{13}^{(c)} c + \nu_{13}^{(f2)} f_2\right) / (f_1 + c + f_2)$$

$$\nu_{13}^{(s)} = \left[\nu_{13}^{(f1)} (f_1 - h) + \nu_{13}^{(c)} c + \nu_{13}^{(f2)} f_2\right] / (f_1 - h + c + f_2)$$

*

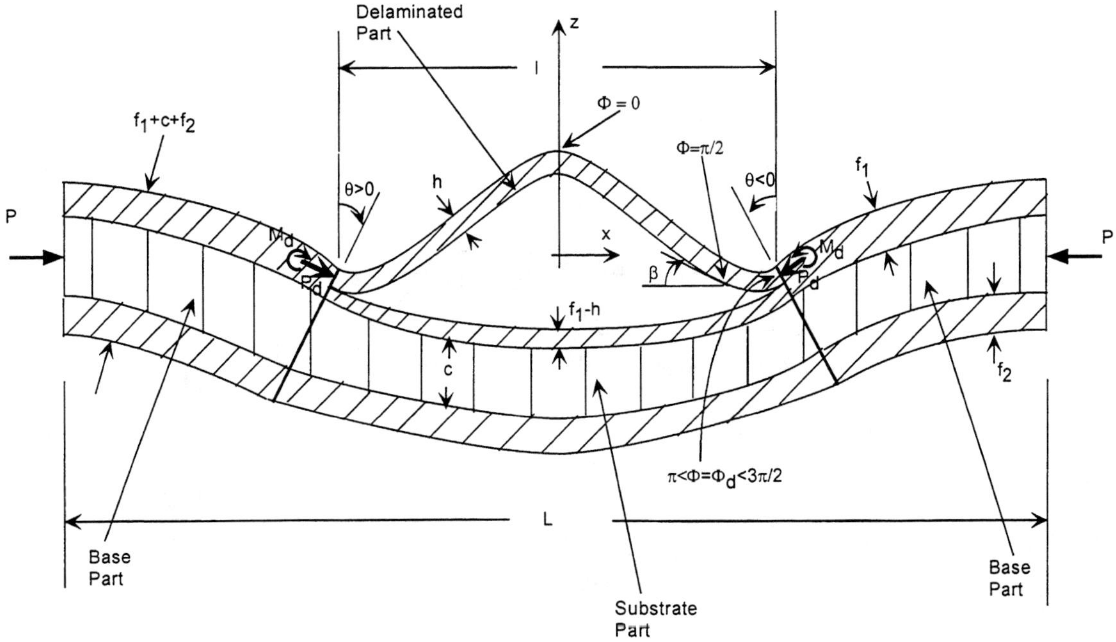

Figure 1. A sandwich beam with a delamination on the top skin/face sheet.

*

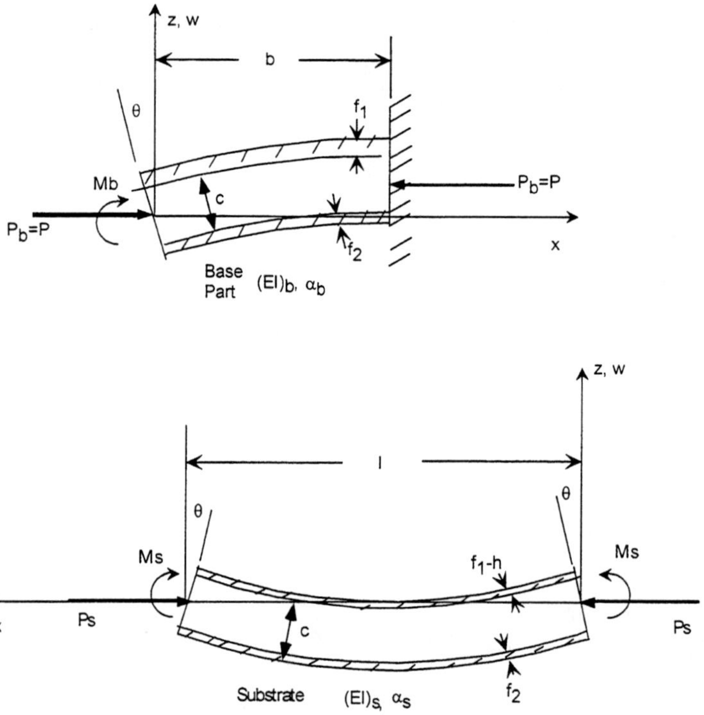

Figure 2. Substrate and Base parts.
*

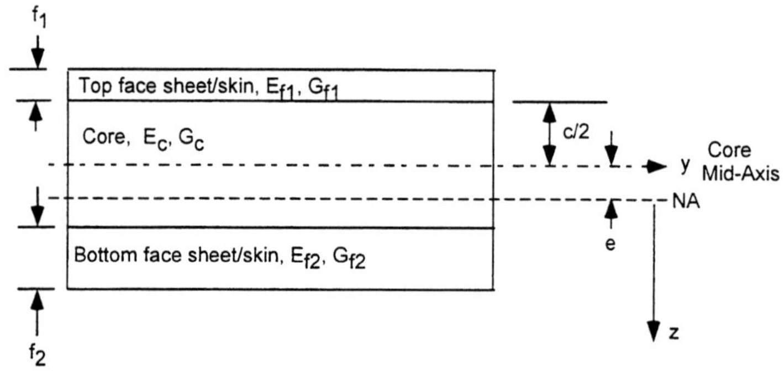

Figure 3. A section of an unsymmetric sandwich beam.

AMD-Vol. 235, Thick Composites for Load Bearing Structures
ASME 1999

THREE DIMENSIONAL EFFECTS IN COMPOSITE PLATES

Santosh Prabhu and John Lambros *

Department of Mechanical Engineering
University of Delaware
Newark, DE 19716, USA

ABSTRACT

In this study, a finite element investigation of the three dimensional nature of stress fields in the near tip region of a cracked orthotropic plate was conducted. Two and three dimensional finite element analyses were used to investigate the relative extent of regions of three dimensional to two dimensional (plane stress or plane strain) deformation in the cracked plate. The material properties used in the simulations corresponded to those of a graphite/epoxy composite. A three point bend loading geometry, with the fiber directions either parallel or perpendicular to the crack, was simulated. In analogy to isotropic materials, it was observed that a plane stress K-dominant region does not arise arbitrarily close to the crack tip because of the existence of a three dimensional zone. However, it was seen that the shape and the size of this three dimensional zone in the cracked composite plate is substantially different from that of an isotropic plate, and depends intimately on material properties. For a crack parallel to the fiber direction the three dimensional zone extends to 0.46h (h = specimen thickness) ahead of the crack tip but only to 0.27h at 30°. Fibers perpendicular to the crack produce a highly elongated three dimensional zone in the direction of the fibers (up to 0.78h). The zone is also sensitive to the variations in the Poisson's ratio's of the orthotropic solid.

INTRODUCTION

Many of the existing failure theories are based on the concepts of critical stress intensity factor, K_{IC}, critical strain energy rate, G_C, or J-integral, J_C. These quantities by their very definition are two dimensional in nature and have to be determined experimentally. In recent years, it has increasingly been recognized that the stress and deformation fields surrounding a crack tip are highly three dimensional in nature (Rosakis and Ravi-Chandar, 1986; Narasimhan and Rosakis, 1988; Krishnaswamy et al. 1991). This fact must be taken into account in order to extract meaningful values of fracture parameters, such as K_I, G and J, from experiments.

Using an edge cracked three point bend geometry Rosakis and Ravi-Chandar (1986) experimentally studied the extent of three dimensionality of crack tip stress fields in isotropic materials. Their results indicated that plane stress conditions prevail at distances from the crack tip greater than half the specimen thickness. In detailed a numerical study Krishnaswamy et al. (1991) investigated the extent of dominance of the mode I asymptotic field surrounding a dynamically loaded stationary crack. Their results also confirmed the existence of a three dimensional stress field at distances up to half the specimen thickness away from the crack tip.

Three dimensional deformation in composite materials has also been studied in the past, but usually in the context of edge effects in laminated composite materials (Lessard et al., 1996; Davidson et al., 1995; Yang and He, 1994; Choi and Folias, 1994). Lessard et al. (1996) developed a three dimensional finite element analysis model to analyze the free-edge effect in a symmetric cross-ply composite laminate of finite dimensions. Davidson et al. (1995) investigated the effects of stacking sequence on the energy release rate in laminated composite end-notched flexure (ENF) test specimens. They noted that the two primary sources of three dimensional effects in multidirectional ENF geometries are the bending-stretching or bending-twisting behavior of the sublaminates and the finite width of the specimen. Yang and He (1994) analyzed the interlaminar free edge stresses and free edge delamination of composite laminate. They studied the strain energy release rate in a [±25/90] graphite-epoxy rectangular laminate subjected to tensile strain. Their results showed a non-uniform distribution of mode I strain energy release rate along the boundary of the delamination front, which fluctuates to a maximum value when approaching both ends at the free edge of the laminate. Choi and Folias (1994) studied the three dimensional stress field in a

* Author to whom correspondence should be addressed.

laminated composite plate that was weakened by the presence of a hole and had been subjected to a uniform displacement along the horizontal direction. Their analysis showed that the interlaminar normal stress depends on the plate thickness but the interlaminar shear stress did not have any appreciable thickness dependency. From these studies we can conclude that the parameters of interest that may affect the development of three dimensional effects around a crack tip are: specimen thickness, crack length, material anisotropy, stacking sequence, loading mixity. The first three will be addressed in the present work.

With increased use of composite materials in engineering applications there is a renewed interest in determining the static and dynamic fracture toughness of composite materials. Liu et al. (1997) have utilized the CGS (Rosakis, 1993) interferometric technique to measure the quasi-static fracture toughness of a fiber reinforced polymer matrix composite. Lambros and Rosakis (1997) utilized the same optical technique and loading geometry under impact loading conditions. In both these studies, the region of three dimensionality surrounding the crack tip was taken to be that of an isotropic material i.e. 0.5 times the plate thickness. This, however, need not be the case for an anisotropic material. In the present study, the finite element method is used to study the three dimensionality of crack tip fields in a transversely isotropic material (meant to simulate a unidirectional fiber reinforced composite at a continuum scale). The results will allow a more accurate analysis of experimental measurements made in the fracture of unidirectional composites. The effect of crack length and plate thickness was investigated. The direction of orthotropy was chosen either parallel to or normal to the crack line (i.e. 0° or 90° fibers). Finally, the effect of varying material properties was also studied.

TWO DIMENSIONAL ANALYSIS

Problem formulation

The three point bend specimen shown in Fig. 1 was modeled using two dimensional plane stress finite elements. This geometry was selected to be the same as that used in the experimental study of Liu et al. (1997), Lambros and Rosakis (1997) and Prabhu and Lambros (1999). Because of symmetry only one-half of the plate was discretized as shown in Fig. 2(a). An x_1-x_2 rectangular Cartesian coordinate system is centered at the crack tip. With reference to this coordinate system the following boundary conditions are imposed on the plate

$$F(W-a,0) = -P/2,$$
$$u_1(-a, L/2) = 0,$$

where F is the nodal force used, P is the total applied load, $u_\alpha(x_1,x_2)$ (α=1,2) are the displacements along the x_α directions and dimensions W, a, L are defined in Fig. 1. In addition, the symmetry condition applied was

$$u_2(x_1,x_2)=0, \qquad 0 \le x_1 \le W\text{-}a.$$

Table 1 – Baseline elastic constants of the graphite/epoxy composite laminate

STIFFNESS PROPERTIES
E_{11} = 150 GPa
$E_{22} = E_{33}$ = 11.6 GPa
$\nu_{12} = \nu_{13}$ = 0.36
ν_{23} = 0.187
μ_{12} = 3.5 GPa

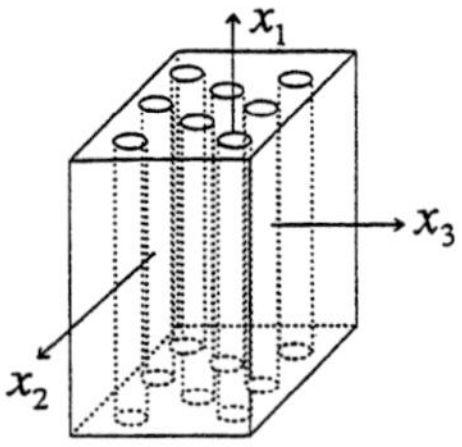

The remainder of the plate was held traction free. Since in the present study we are only interested in capturing the details of the stress field in the vicinity of a stationary crack, a radially focused mesh centered at the crack tip was used in the near tip region. The elements used were eight-noded quadrilateral elements with the exception of the ring of elements closest to the crack tip which were six-noded triangles. The resulting mesh, therefore, had a total of 408 elements, 270 of which were in the focused region. This yielded a total of 1281 nodes. To simulate a unidirectional fiber reinforced composite plate a transversely isotropic constitutive law was employed. The baseline material properties selected (see Table 1) were the same as those quoted in Lambros and Rosakis (1997). All analysis is actually performed for homogeneous orthotropic plates, with the conclusions extended to unidirectional, fiber-reinforced composite plates under the assumption that the fiber size is small compared to the plate thickness. The ABAQUS (1998) finite element software package was used throughout this work.

Results

Referring to the coordinate system shown in Fig. 1 it can be shown that the opening stress, σ_{22}, is asymptotically given by (Sih et al., 1965; Sih and Liebowitz, 1968; Liu et al., 1997)

$$\sigma_{22} = \frac{K_I}{2\sqrt{2\pi}}\left\{\alpha_2(b_{ij})r_1^{-1/2}\cos\frac{\theta_1}{2} - \beta_2(b_{ij})r_2^{-1/2}\cos\frac{\theta_2}{2}\right\}, \quad (1)$$

where $r_\alpha = \sqrt{x_1^2 + p_\alpha^2 x_2^2}$, $\theta_\alpha = \tan^{-1}(p_\alpha x_2/x_1)$ are scaled polar coordinates and p_α, α_2 and β_2 are functions of material compliance b_{ij} and are explicitly given in Liu et al. (1997). K_I is the mode I stress intensity factor. Equation (1) is similar to the analogous result for isotropic materials. It exhibits the usual $1/\sqrt{r}$ singularity as the crack tip is approached. It differs from the isotropic result in the functions α_2 and β_2, which depend only on material properties. Figure 3(a) shows, on a logarithmic scale, the variation of normalized opening stress versus normalized distance ahead of the crack tip (i.e. for $\theta = 0°$) for the case in which the direction of orthotropy (i.e. fiber direction) is parallel to the x_1 axis (i.e. the crack line). A large portion of the curve is linear with a slope of -1/2, clearly indicating the range over which Eq. (1) (solid line in Fig. 3(a)) is valid. Figure 3(b) shows a similar result for the case in which the fiber direction is perpendicular to the crack. The decreased range of validity of (1) is because of the shielding effect provided by the fibers.

THREE DIMENSIONAL ANALYSIS

Problem formulation

The problem of interest in this section is again the three point bend geometry of Fig. 1, but with a plate thickness of h. Since only symmetric loading will be considered we need only discretize one-quarter of the plate (half width and half thickness). The in-plane mesh geometry is the same as that of the two dimensional problem (Fig. 2(a)). In the third direction (x_3) seven layers of elements are fitted through half the thickness of the specimen. The layers become thinner as the free surface is approached in order to capture the expected larger stress variation near the free surface of the plate. Specifically, the element adjoining the mid-surface of the plate is three times thicker than the element nearest the free surface. The through thickness mesh is shown in Fig. 2(b). The baseline material properties used were the same as those in the two dimensional analysis. The same boundary conditions were also imposed with the addition of a symmetry condition on the mid-plane of the specimen,

$$u_3(x_1,x_2,0) = 0.$$

The resulting mesh had a total of 1890 elements in the focused region. This is substantially higher than the number used in the studies of Krishnaswamy et al. (1991) and Lee and Rosakis (1993).

Measure of three dimensionality

In the work of Krishnaswamy et al. (1991) and Lee and Rosakis (1997) the importance of three dimensional effects was monitored through the parameter $\sigma_{33}/\nu(\sigma_{11}+\sigma_{22})$ where σ_{ij} are the components of the stress tensor and ν is the Poisson's ratio (for the isotropic material). This parameter, referred to as the degree of plane strain, becomes 0 under conditions of plane stress and 1 under conditions of plane strain. Using this parameter it was seen (Krishnaswamy et al., 1991) that for the case of a cracked isotropic plate, three dimensional conditions prevailed up to a distance of 0.5h surrounding the crack tip.

In the case of an anisotropic material, the degree of plane strain would no longer be given by the above parameter. In general, it would depend on the details of the anisotropic constitutive law chosen. For the case of an orthotropic material the stress-strain relationship can be expressed as:

$$\begin{Bmatrix}\sigma_{11}\\ \sigma_{22}\\ \sigma_{33}\\ \sigma_{31}\\ \sigma_{32}\\ \sigma_{12}\end{Bmatrix} = \begin{bmatrix} \frac{1}{E_{11}} & -\frac{\nu_{12}}{E_{11}} & -\frac{\nu_{13}}{E_{11}} & 0 & 0 & 0\\ -\frac{\nu_{21}}{E_{22}} & \frac{1}{E_{22}} & -\frac{\nu_{23}}{E_{22}} & 0 & 0 & 0\\ -\frac{\nu_{31}}{E_{33}} & -\frac{\nu_{32}}{E_{33}} & \frac{1}{E_{33}} & 0 & 0 & 0\\ 0 & 0 & 0 & \frac{1}{2\mu_{31}} & 0 & 0\\ 0 & 0 & 0 & 0 & \frac{1}{2\mu_{32}} & 0\\ 0 & 0 & 0 & 0 & 0 & \frac{1}{2\mu_{12}} \end{bmatrix} \begin{Bmatrix}\varepsilon_{11}\\ \varepsilon_{22}\\ \varepsilon_{33}\\ \varepsilon_{31}\\ \varepsilon_{32}\\ \varepsilon_{12}\end{Bmatrix}, \qquad (2)$$

where σ_{ij} and ε_{ij} are the stress and the strain tensor components, respectively. The E's, ν's and μ's are the Young's moduli, Poisson's ratios and shear moduli of the orthotropic material, respectively. For a plane strain situation $\varepsilon_{33} = 0$, where x_3 axis represents the through thickness direction. Therefore, from Eq. (2), for plane strain,

$$\varepsilon_{33} = -\frac{\nu_{13}}{E_{11}}\sigma_{11} - \frac{\nu_{23}}{E_{22}}\sigma_{22} + \frac{1}{E_{33}}\sigma_{33} = 0. \qquad (3)$$

For plane stress, $\sigma_{33} = 0$. We therefore define the following quantity λ as the degree of plane strain for the unidirectional composite,

$$\lambda = \frac{\sigma_{33}}{E_{33}\left(\frac{\nu_{13}}{E_{11}}\sigma_{11} + \frac{\nu_{23}}{E_{22}}\sigma_{22}\right)} \qquad (4)$$

It can been seen that for a plane strain situation, $\lambda=1$ and for a plane stress situation, $\lambda=0$. Note that this is the case only when the material and the loading axes coincide.

Results and Discussion

The focus of the present study is complementary to the experimental work in Prabhu and Lambros (1999). Thus, its instructive to use the present results to investigate the region of K-dominance in the three point bend orthotropic specimen. (A region of K-dominance is defined as that in which the stress field is well approximated by the leading term of the asymptotic expression for stress, i.e. Eq. (1).) It has been shown that for isotropic materials a K-dominant region does not arise arbitrarily close to the crack tip because of the presence of three dimensional and process zone effects. In addition, further away higher order effects become important. Therefore, the K-dominant region surrounding the crack tip is an annular region as shown schematically in Fig. 4. The two dimensional finite element analysis assists in determining the maximum extent of the K-dominant field. When the finite element results shown as squares in Fig. 3(a,b) deviate from those predicted by Eq. (1) (solid line in Fig. 3(a,b)) as the radial distance increases, the singularity dominated region no longer exists and higher order terms are required to adequately describe the stress field. (The deviation in Fig. 3(a,b) very close to the crack tip is mostly due to numerical discretization errors.) The plane stress solution thus provides us with an outer bound of the region where the singularity dominated solution is valid. In an experimental setting, we can obtain meaningful values of the stress intensity factor K_I if this outer bound is substantially larger than the inner bound of the K-dominant region.

The inner bound of the K-dominant region will be dictated by the larger of the radii of the process, yield or three dimensional region. For the case of the composite the largest is expected to be the three dimensional region. Fig. 5 shows a three dimensional contour plot of the degree of plane strain, λ for a transversely isotropic plate of thickness h=6.35 mm, length L=305.4 mm, width, W=76.2 mm and crack length to width ratio, a/W=0.2. The baseline material properties (Table 1) were used. In this material set the fiber direction (i.e. the largest value of E) is along the crack. The a/W=0. 2 was selected because it has been shown to provide a substantial region of K-dominance in isotropic materials (Narasimhan and Rosakis, 1990; Krishnaswamy et al., 1991) The top surface visible in Fig. 5 corresponds to the mid-plane of the plate. The first contour indicates a $\lambda \approx 0.1$ i.e. a 10% variation from the region of two dimensional plane stress and can be considered as the inner boundary of the K-dominant region. As can be seen from Fig. 5, this contour has an irregular shape: it extends to 0.46*h* (h=thickness) directly ahead of the crack tip, only to about 0.27h at an angle of 30° and up to 0.6h at 90° to the crack. The shape of this three dimensional region is substantially different from that of the isotropic and bimaterial cases (Krishnaswamy and Rosakis, 1991; Lee and Rosakis, 1993). As the outer surface of the plate is approached the deviation from plane stress conditions becomes less (a consequence of the fact that $\sigma_{33}=0$ on the free surface). At distances very close to the crack tip, r/h<0.01, λ approaches a value of 1. This indicates that very close to the crack tip, a plane strain region exists.

Several factors may affect the development of the three dimensional stress state ahead of the crack tip. These include load, thickness, crack length, fiber orientation and material properties. It was observed that the shape and size of the three dimensional zone does not change with the applied load. As the load changes, the stress values change proportionally so that the value of λ remains unchanged.

It was also seen that the size of the three dimensional zone scales with thickness whereas its shape remains unchanged. For example, if the thickness of the plate is doubled, the size of the three dimensional region also doubles but it shape remains unchanged. This effect was also observed in the isotropic and bimaterial cases (Krishnaswamy et al., 1991; Lee and Rosakis, 1993). In addition, for the case in which the crack length was increased to a/W=0.4 the results remain identical to those presented in Fig. 5(a) (i.e. for the case a/W=0.2). This would remain true as long as the crack does not interact with the boundaries of the plate.

Figure 6 shows the contours of λ for a plate of thickness 6.35 mm with a/W=0.2 but with fibers perpendicular to the crack (along the x_2-direction). The baseline properties of Table 1 have been used, but with the values of E_{11} and E_{22} (E_{33}) interchanged. As is clear, the extent of dominance of the three dimensional effects is substantially different in this case (compare with Fig. 5(a)). The region of three dimensionality extends to only about 0.5h directly ahead of the crack, has a minimum of 0.28h at 40° but at 90° to the circle extends up to 0.78h. This has important implications on the extent of K-dominance as a function of angle around the crack tip.

The effect of "direction" and "amount" of orthotropy is shown Fig. 7 (a,b,c). The "direction" of orthotropy is measured by the ratio $\beta = E_{11}/E_{22}$ where E_{11} and E_{22} are the Young's moduli in the 1 and 2 directions, respectively. If $\beta > 1$ the composite fibers, i.e. direction of orthotropy, may be considered as being parallel to the crack and if $\beta < 1$ the fibers may be considered perpendicular to the crack. The "amount" of orthotropy is larger, the further β deviates from a value of 1. Figure 7(a,b,c) present a sequence of three contour plots illustrating effect of degree of orthotropy on the shape of the three dimensional zone as measured by λ. The values of β presented in this plot are 1/5, 1, 5, respectively. Varying values of β were obtained by changing the values of E_{11}, E_{22} and E_{33} from the baseline values of Table 1. In Fig. 7(a,b,c), only the midplane of the specimen is illustrated. (The through thickness variation is similar to that of Fig. 5) As expected for β=1 (Fig. 7(b)) the shape of the three dimensional zone is identical to that of the isotropic material, previously studied by Krishnaswamy et al (1991). It is almost circular in shape and extends to approximately 0.5*h* around the crack tip. As β increases above 1, we see that the three dimensional zone becomes elongated in the direction of orthotropy, i.e. in the direction of the crack (Fig. 7(c)). When β is made lower than 1 (Fig. 7(a)), the three dimensional zone again becomes elongated in the direction perpendicular to the direction of orthotropy, i.e. in this case perpendicular to the direction of the crack line. Figures 5 and 6, which essentially correspond to β = 12.93 and 1/12.93 respectively, also show an agreement with this observed trend.

The extent of the three dimensional zone is also significantly affected by the variations in Poisson's ratios, especially ν_{23}. For the case in which the baseline properties are modified by having $\nu_{12} = \nu_{23} = \nu_{13} = 0.36$, the three dimensional zone considerably elongates in a direction perpendicular to the direction of orthotropy, as seen in Fig. 8. The strong dependence on Poisson's ratio's, ν_{23} and ν_{13}, is expected, since in this geometry these Poisson's ratio's are mostly responsible for the development of the through thickness deformation. A much smaller influence of the in-plane Poisson's ratio, ν_{12}, was seen.

CONCLUSIONS

In this paper we used the finite element method as a means of evaluating the significance of three dimensional effects in an edge cracked orthotropic plate subject to three point bend loading. Very close to the crack tip plane strain conditions are approached and further away plane stress conditions prevail. In a large region surrounding the crack tip, however, the stresses are three dimensional in nature. The shape of the three dimensional region differs dramatically from that of the homogeneous and the bimaterial cases. It depends on plate thickness and fiber orientation but is independent of crack length. For a crack parallel to the fiber direction the three dimensional zone extends up to 0.46*h* ahead of the crack tip but only 0.27*h* at 30°. Further, fibers perpendicular to the crack produce a highly elongated three dimensional zone in the direction of the fibers (up to 0.78*h*). However, the specific size of the three dimensional zone is seen to depend on the amount of orthotropy, as quantified by the ratio $\beta = E_{11}/E_{22}$, and on the Poisson's ratios, especially ν_{23}.

ACKNOWLEDGEMENTS

We would like to acknowledge the generous support of the Department of Mechanical Engineering of the University of Delaware and the University of Delaware Research Foundation (LTR 19960415).

REFERENCES

1. ABAQUS/Standard Users Manual, Version 5.7, 1998, Hibbit, Karlsson & Sorensen Inc., Pawtucket, RI.
2. Choi, Y.K., and Folias, E.S., "Failure Mechanics in Advanced Polymeric Composites" ASME 1994, AMD-Vol. 196, pp. 31-42.
3. Davidson, B.D., Kruger, R. and Konig, M., 1995, " Three Dimensional Anaylsis and Resulting Recommendations for Unidirectional and Multidirectional Single Leg Bend Specimens," Composite Science and Technology, Vol. 54, pp 385-394.
4. Krishnaswamy, S., Rosakis, A.J., and Ravichandran, G., 1991, "On the Extent of Dominance of Asymptotic Elastodynamic Crack-tip Fields: Part II- Numerical Investigation of Three Dimensional and Transient Effects," Journal of Applied Mechanics, Vol. 58, pp. 95-103.
5. Lambros, J., and Rosakis, A.J., 1993, "Dynamic Crack Initiation and Growth in Thick Unidirectional Graphite/Epoxy Plates," Composites Science and Technology, Vol. 57, pp. 55-65.
6. Lee. Y.J., and Rosakis, A.J., 1993, "Interfacial Cracks in Plates: A Three Dimensional Numerical Investigation," International Journal of Solids and Structures, Vol. 30., pp. 3139-3158.
7. Lessard, L.B., Schmidt, A.S., and Shokrieh, M.M., 1996, "Three-dimensional Stress Analysis of Free Edge Effects in a Simple Cross-ply Laminate," International Journal of Solids and Structures, Vol. 33., pp. 2243-2259.
8. Liu, C., Rosakis, A.J., Stout, M.G., and Ellis, R., 1997, "On the Application of CGS Interferometry to the Study of Fiber Reinforced Composites," submitted to International Journal of Fracture.
9. Narasimhan, R. and Rosakis, A.J., 1988, "Three Dimensional Effects Near a Crack Tip in a Ductile Three Point Bend Specimen. Part I: Numerical Analysis," Journal of Applied Mechanics, Vol. 57, pp. 607-671.
10. Prabhu, S., and Lambros, J., 1999, "On the Role of Orthotropy in the Three Dimensional Stress and Deformation Crack Tip Fields," under preparation.
11. Rosakis, A.J. and Ravi-Chandar, K., 1986, International Journal of Solids and Structures, Vol. 22, pp. 121-134.
12. Rosakis, A.J., 1993, "Two Optical Techniques Sensitive to Gradients of Optical Path Difference: The Method Of Caustics and the Coherent Gradient Sensor (CGS)," Experimental Techniques in Fracture (Ed. J.S. Epstein), VCH Publishers Inc., New York, pp. 327-425.
13. Sih, G., Paris, P.C., Irwin, G.R., 1965, "On Cracks in Rectilinearly Anisotropic Bodies," International Journal of Fracture, Vol. 1. Pp. 189-203.
14. Sih, G.C., and Liebowitz, H., 1968, "Mathematical Theories of Brittle Fracture," Fracture, Academic Press, New York, pp. 67-189.
15. Yang, H.T.Y. and He, C.C., 1994, Three-dimensional Finite Element Analysis of Free Edge Stresses and Delamination of Composite Laminates, Vol. 28, pp. 1394-1412.

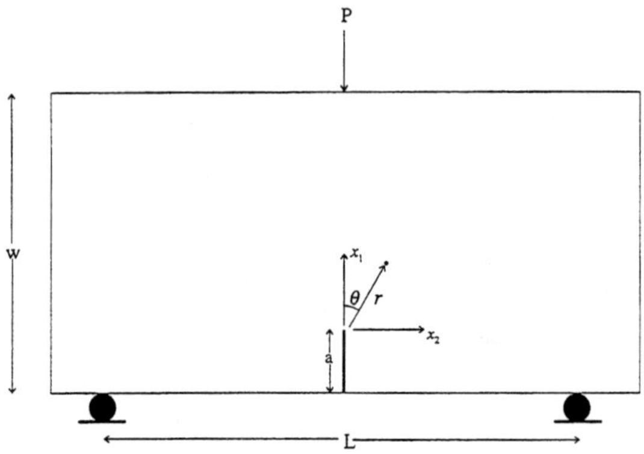

Fig. 1 – Three point bend geometry and definition of the coordinate axes.

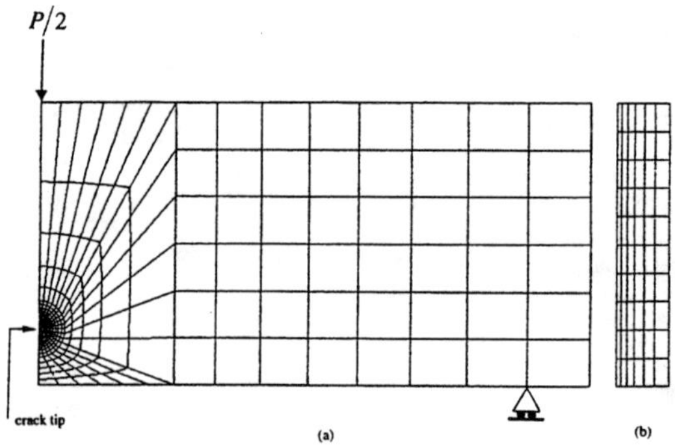

Fig. 2 – Finite element mesh for the two dimensional (a) and three dimensional (a,b) analyses.

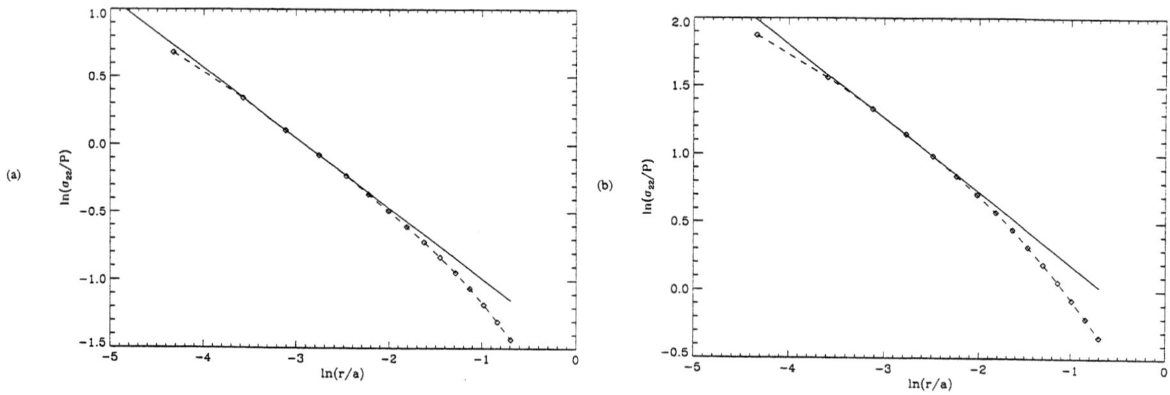

Fig. 3 – Logarithmic plots of opening stress ahead of the crack (θ=0°) under two dimensional plane stress conditions for the case of (a) fibers parallel to the crack and (b) fibers perpendicular to the crack. In each case, solid line represents the result of Eq. (1).

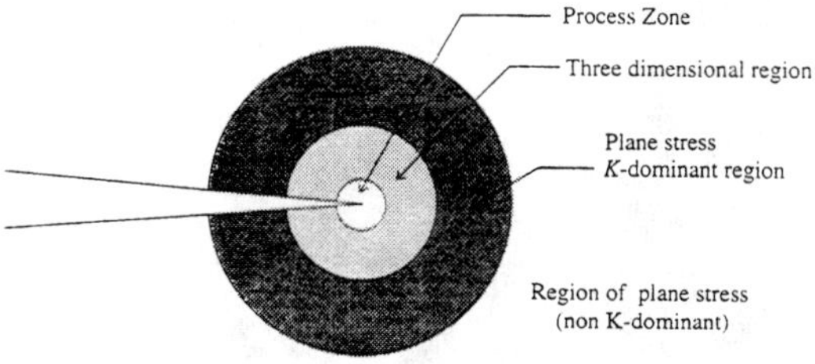

Fig. 4 – Schematic of deformation regions around a crack tip.

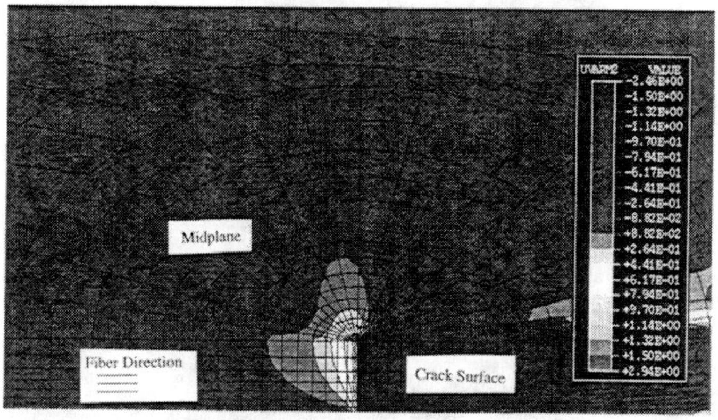

Fig. 5 – Variation of the degree of plane strain, λ, for a three point bend transversely isotropic plate with a/W=0.2, h=6.35 mm and fibers parallel to the crack line.

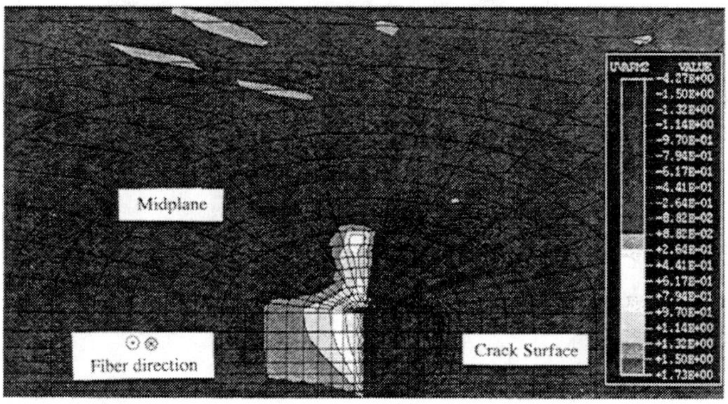

Fig. 6 - Variation of the degree of plane strain, λ, for a three point bend transversely isotropic plate with a/W=0.2, h=6.35 mm and fibers perpendicular to the crack line.

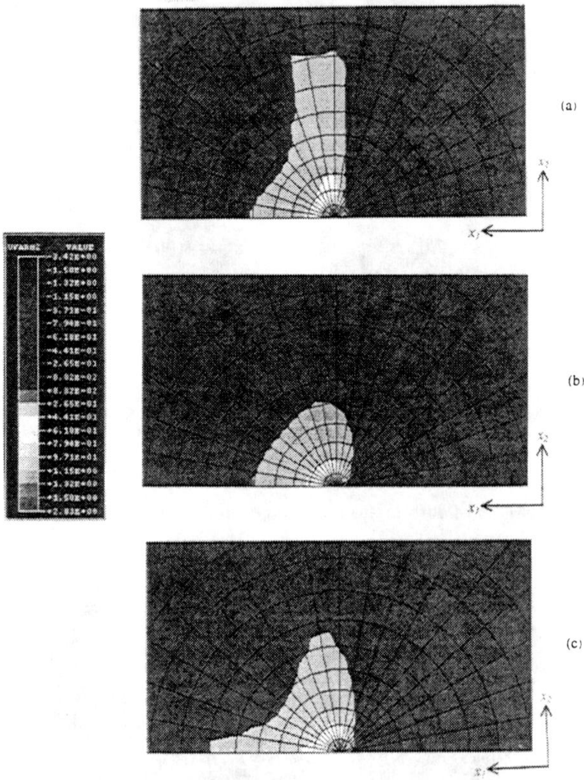

Fig. 7 – Contour plots of degree of plain strain, λ, for (a) β=1/5, (b) β=1, and (c) β=5 for a plate under three point bend loading. Midplane (x_3=0) of the specimen is illustrated.

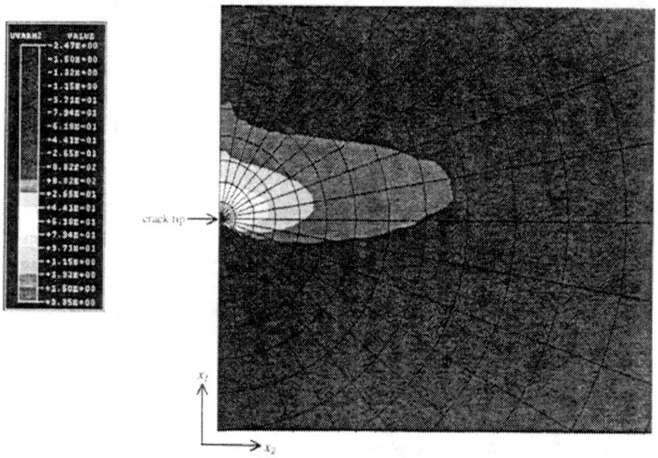

Fig. 8 – Contour plot of degree of plane strain λ for the case of $\nu_{12}= \nu_{13}=\nu_{23}=0.36$. Midplane of the specimen (x_3=0) is illustrated.

AMD-Vol. 235, Thick Composites for Load Bearing Structures
ASME 1999

EQUIVALENT COMPOSITE LAMINATED BEAMS WITH VARIOUS ELASTIC CONSTITUTIVE MODELS

Izhak Sheinman and **Yeoshua Frostig**

Faculty of Civil Engineering
Technion – Israel Institute of Technology

ABSTRACT

Equivalent one-dimensional constitutive models of composite laminated beams with shear deformation are derived from the classical laminate two-dimensional using first-order shear deformable theory. The present cylindrical bending constitutive models can be used - with much greater accuracy than their well known plane-strain and plane-stress counterparts – as upper and lower bounds, to one of which the behavior tends depending on the width-to-length ratio; this aspect was investigated and results are presented.

INTRODUCTION

A two-dimensional model with two opposite free edges in the longitudinal direction and supported along the other pair of edges, can be replaced by an equivalent one-dimensional model, namely a beam element. Obviously, the advantage of this equivalent element is that it involves only two dependent variables (the longitudinal and the vertical displacements). However, for a composite laminate, compliance with the boundary conditions in the width direction is questionable.

In the case of composite laminated panels two well known versions of the constitutive model are used - plane-strain and plane-stress (for large and small width-to-length ratios, respectively). They are almost the same for the isotropic case and are used only as upper and lower bounds. However due to the significant gap between them, more accurate models are called for.

Whitney's (1969) cylindrical bending theory (referred to here by **cylW.**), made for some improvement over the plane-strain version by narrowing the above gap. Further improvement with regard to the plane-stress version (referred to **cylN.**) is proposed below.

FORMULATION

A beam element represented by a one-dimensional model (in the x-z plane) with only two dependent variables (displacement in the longitudinal (u) and vertical (w) directions) and the general force-strain relations can be written as:

$$N_{xx} = a\bar{\varepsilon}_{xx} + \bar{b}\chi_{xx}$$
$$M_{xx} = \bar{b}\bar{\varepsilon}_{xx} + d\chi_{xx} \qquad (1)$$

Here x – is the coordinate of the reference surface of the beam, N_{xx} and M_{xx} are the axial membrane force and bending moment, $\bar{\varepsilon}_{xx}$ and χ_{xx} the strain of the reference surface and the change of curvature, respectively.

For classical isotropic beams $\bar{b} = 0$, and a and d are straightforward axial (EA) and bending (EI) stiffnesses. By contrast, for a composite laminated beam element with an arbitrary stacking combination and orientation, a, b and d are no longer uniquely determined and their values depend on the applied equivalent anisotropic theory.

Different equivalent one-dimensional models (represented by a, $\bar{b}$ and d of eq. (1)), can be derived from the classical two-dimensional laminate theory, for which the stress-strain relation for each lamina (j) is given by:

$$\begin{Bmatrix} \sigma_{xx} \\ \sigma_{yy} \\ \tau_{xy} \end{Bmatrix}_j = \begin{bmatrix} \bar{Q}_{11} & \bar{Q}_{12} & \bar{Q}_{13} \\ \bar{Q}_{12} & \bar{Q}_{22} & \bar{Q}_{23} \\ \bar{Q}_{13} & \bar{Q}_{23} & \bar{Q}_{33} \end{bmatrix}_j \begin{Bmatrix} \varepsilon_{xx} \\ \varepsilon_{yy} \\ \gamma_{xy} \end{Bmatrix}_j \qquad (2)$$

where

$$\begin{Bmatrix} \varepsilon_{xx} \\ \varepsilon_{yy} \\ \gamma_{xy} \end{Bmatrix}_j = \begin{Bmatrix} \bar{\varepsilon}_{xx} \\ \bar{\varepsilon}_{yy} \\ \bar{\gamma}_{xy} \end{Bmatrix} + z_j \begin{Bmatrix} \chi_{xx} \\ \chi_{yy} \\ 2\chi_{xy} \end{Bmatrix} \qquad (3)$$

$\{\bar{\varepsilon}\}$ and $\{\chi\}$ are the strain-of-the-reference-surface and change-of-curvature vectors, respectively, $\bar{Q}_{ij}$ the transformed stiffness coefficients of the laminate.

Integration over the thickness yields the following force and moment resultant of the two-dimensional laminate:

$$\begin{Bmatrix} N \\ M \end{Bmatrix} = \begin{bmatrix} A & B \\ B & D \end{bmatrix} \begin{Bmatrix} \bar{\varepsilon} \\ \chi \end{Bmatrix} \qquad (4)$$

where $\{N\} = \{N_{xx}, N_{yy}, N_{xy}\}$ is the membrane force vector, $\{M\} = \{M_{xx}, M_{yy}, M_{xy}\}$ the bending moment vector. A, B and D are matrices of order 3*3, given by:

$$(A_{ij}, B_{ij}, D_{ij}) = \int_z \bar{Q}_{ij} (1, z, z^2) dz \qquad (5)$$

In the following, two equivalent one-dimensional models, cylindrical of wide and narrow beams with shear deformation are considered. For plane- strain and plane stress see Whitney (1987).

Cylindrical Bending – Wide Beam (CylW)

This model is more accurate than the plane-strain version and takes into account the fiber orientation. It is derived assuming that the displacement vector of the two-dimensional model – is a function of the longitudinal coordinate alone (see Whitney (1969), (1987)):

$$
\begin{aligned}
u(x, y, z) &= \bar{u}(x) + z\psi(x) \\
v(x, y, z) &= \bar{v}(x) \qquad (6)\\
w(x, y, z) &= \bar{w}(x)
\end{aligned}
$$

u , v and w are the displacement functions in the x , y and z-directions of the two-dimensional model.

The appropriate kinematic relations based on the above assumption (eqs. (6)) are:

$$
\begin{aligned}
\bar{\varepsilon}_{xx} &= \bar{u}_{,x} + \frac{1}{2}\bar{w}_{,x}^2 \\
\bar{\varepsilon}_{yy} &= \bar{\gamma}_{yz} = \chi_{yy} = \chi_{xy} = 0 \\
\bar{\gamma}_{xy} &= \bar{v}_{,x} \qquad (7)\\
\bar{\gamma}_{xz} &= \bar{w}_{,x} + \psi \\
\chi_{xx} &= \psi_{,x}
\end{aligned}
$$

The variational principle (based on eqns. (7)) yields the following equilibrium equations:

$$
\begin{aligned}
& N_{xx,x} + q_{xx} = 0 \\
& N_{xy,x} = 0 \qquad (8)\\
& Q_{xz,x} + \left(N_{xx} w_{,x}\right)_{,x} + q_{zz} = 0 \\
& M_{xx,x} - Q_{xz} + m = 0
\end{aligned}
$$

q_{xx} , q_{zz} and m being the external applied longitudinal loads, vertical loads and moments, respectively. Note that the equilibrium conditions of the equivalent one-dimensional model (eqns. 8), contain the additional equation:

$$N_{xy,x} = 0 \qquad (9)$$

For the free transverse edges it leads to $N_{xy} = 0$ reads in terms of $\{\bar{\varepsilon}\}$ and $\{\chi\}$ (see eq. (4)):

$$A_{13}\bar{\varepsilon}_{xx} + A_{33}\bar{v}_{,x} + B_{13}\psi_{,x} = 0 \qquad (10)$$

Eliminating $\bar{v}_{,x}$ from eq. (10) and using eqs. (4) and (8), the following relevant stiffness coefficients are obtained:

$$
\begin{aligned}
a &= \left(A_{11}A_{33} - A_{13}^2\right)/ A_{33} \\
\bar{b} &= \left(B_{11}A_{33} - A_{13}B_{13}\right)/ A_{33} \qquad (11)\\
d &= \left(D_{11}A_{33} - B_{13}^2\right)/ A_{33}
\end{aligned}
$$

Note that the stiffness coefficients involve the terms 11, 33 and 13.

Cylindrical Bending – Narrow Beam (CylN)

This model is more accurate than the plane-stress constitutive version. Its derivation is based on inclusion of the effect of fibers orientation with the flexibility coefficients expressed, after the matrix manipulation, as derived by Sheinman (1989) (see also Sheinman and Kardomateas (1997)).

The strain and change of curvature of the one-dimensional model can be written in terms of the longitudinal force and the bending moment as:

$$\bar{\varepsilon}_{xx} = \alpha_1 N_{xx} + \alpha_2 M_{xx}$$
$$\chi_{xx} = \alpha_3 N_{xx} + \alpha_4 M_{xx} \tag{12}$$

Using eqs. (4), writing the strain and change-of-curvature vectors in terms of the membrane-force and bending-moment vectors, and equating to eq. (12) we obtain:

$$\alpha_1 = \left\{\left[I + A^{-1}B\left(D - BA^{-1}B\right)^{-1}B\right]A^{-1}\right\}_{11}$$
$$\alpha_2 = \alpha_3 = \left\{-A^{-1}B\left(D - BA^{-1}B\right)^{-1}\right\}_{11} \tag{13}$$
$$\alpha_4 = \left\{\left(D - BA^{-1}B\right)^{-1}\right\}_{11}$$

The subscript 11 means the first term of the matrix. It is seen that the α's are determined not only by A_{11} , B_{11} and D_{11} but also by A_{ij} , B_{ij} , D_{ij} *(i,j = 1,2,3).*

Finally using eqs. (13), the stiffness coefficients are determined through:

$$a = \alpha_4 / \eta$$
$$\bar{b} = -\alpha_2 / \eta \tag{14}$$
$$d = \alpha_1 / \eta$$
$$\eta = \alpha_1\alpha_4 - \alpha_2\alpha_3$$

RESULTS AND DISCUSSION

For isotropic beams, the plane-strain and plane-stress models are close to each other, so that the differences between the four constitutive models are insignificant. By contrast, non-symmetric composite laminated beams show marked discrepancies caused by the non-vanishing B_{ij} (i,j = 1,3) terms (the larger the terms, the wider the discrepancy). Accordingly, such a laminate with $\alpha^o/-\alpha^o$ lay up was chosen to illustrate the proposed cylindrical bending model approach and its advantage.

The constitutive models are most effectively studied either in the frequency domain or through the buckling behavior. Here, the buckling behavior of a simply-supported laminated (α/–α) beam under an external axial force was chosen as the study case. The data of this case are: material – carbon/epoxy with moduli $E_{11} = 1.4 \cdot 10^{11}\ N/m^2$, $E_{22} = 0.98 \cdot 10^{10}\ N/m^2$, $G_{13} = G_{12} = 0.1 \cdot 10^{10}\ N/m^2$, $\upsilon_{12} = 0.34$, thickness, h=0.00125 m with h_{ply} = 0.000125m - length-to-thickness ratio l/h=10.

The upper and the lower bounds of the normalized buckling loads (P^*_{cr} denoting the buckling load for $\alpha = 0^o$) for a beam with a lay-up of $\alpha^o_5/-\alpha^o_5$ for the various constitutive models determined by the proposed models and compared with the exact results for b/l = 80 and 0.25 obtained by the 2D - plate finite-element COSMOS7 (1984) code, appear in fig. 1 and in fig. 2. It is seen that the buckling load coincides with the **CylW** model for b/l = 80, see fig. 1, and that for the small ratio of b/l, with to the **CylN.** Furthermore, it should also be noted that due to the relatively large value of B_{i3} the bounds lie far below and above the plane-strain and plane-stress curves, respectively.

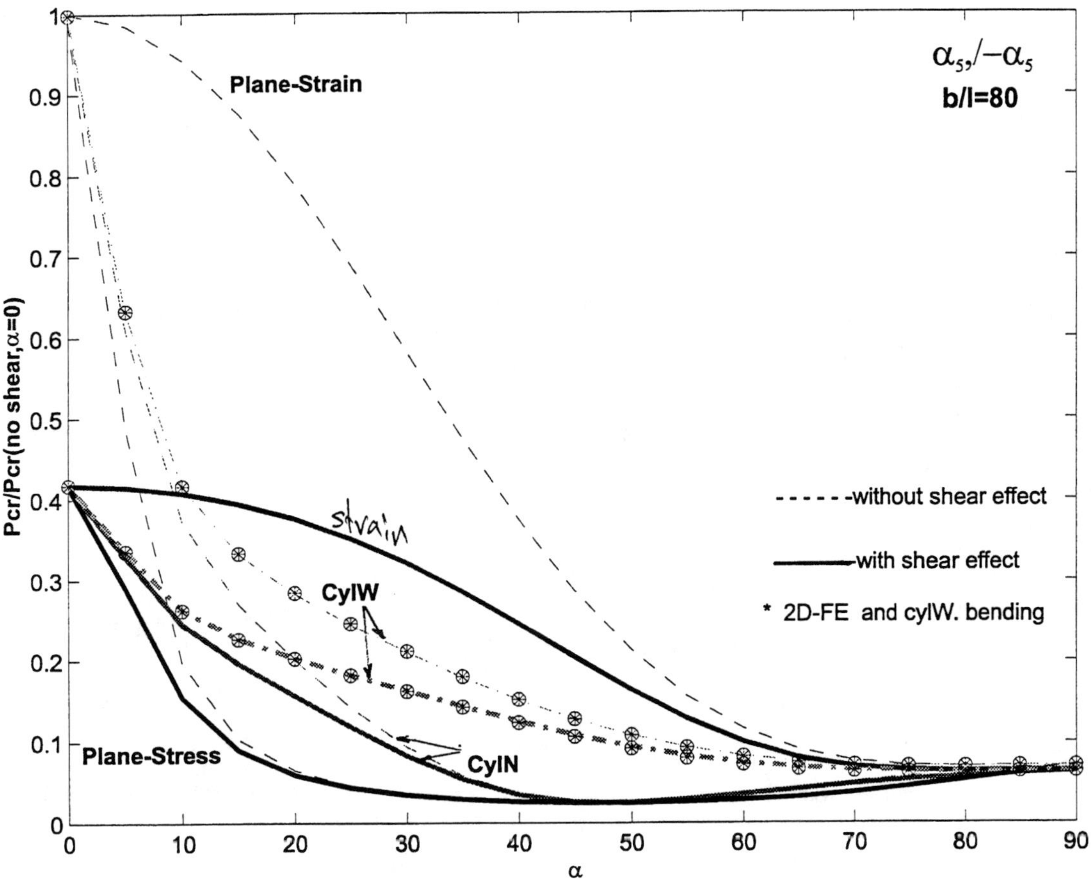

Figure 1: Buckling load versus orientation angle of $\alpha_5/-\alpha_5$ laminate for different constitutive models compared with 2D plate of b/l=80 results (wide beam).

The significant difference between the results with and without shear deformation effect, is due to the low l/h=10 ratio and the high E_{11}/G_{13}=140 ratio, but the same conclusion applies to the constitutive models.

CONCLUSIONS

The well-known plane-strain and plane-stress constitutive models are unsuitable for composite laminated beams. The various cylindrical bending models are much more accurate and appropriate for presentation of laminated stacking combinations and orientations.

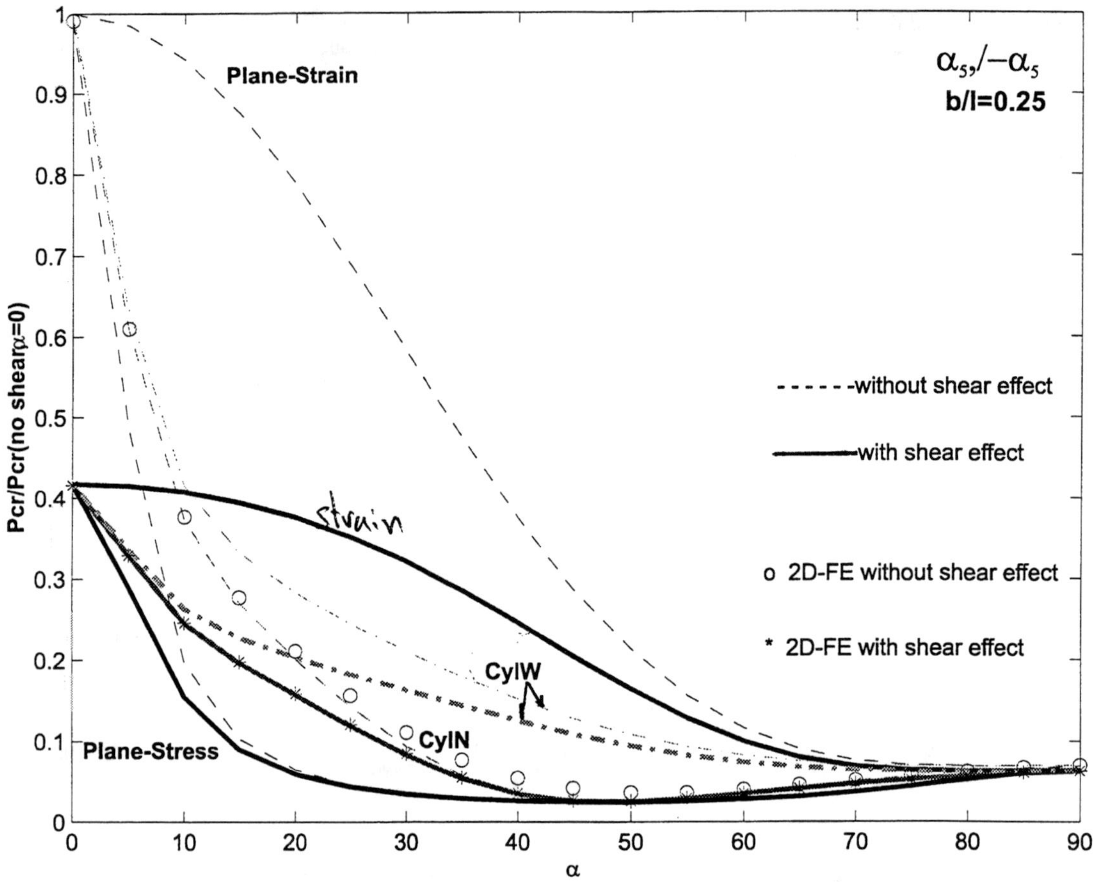

Figure 2: Buckling load versus orientation angle of α_5/-α_5 laminate for different constitutive models compared with 2D plate of b/l=0.25 results (narrow beam).

REFERENCES

COSMOS7, 1984, Structural Research and Analysis Corporation, Santa Monica.

Sheinman, I., 1989, "Cylindrical buckling load of laminated columns", ASCE, Engineering Mechanics, Vol. 115, No. 3, pp. 659-661.

Sheinman, I., and Karadomateas, G.A., 1997, "Energy release rate and stress intensity factors for delaminated composite laminates", Int. J. Solids Structures, Vol. 34, No. 3, pp. 451-459.

Whitney, J.M., 1969, "Cylindrical bending of Unsymmetrically laminated plates", J. of Composite Materials, Vol. 3, pp. 715-719.

Whitney, J.M., 1987, "Structural analysis of laminated anisotropic plates", Technomic publishing Inc.

AMD-Vol. 235, Thick Composites for Load Bearing Structures
ASME 1999

SUBSONIC AND INTERSONIC DYNAMIC CRACK GROWTH IN UNIDIRECTIONAL COMPOSITES

Demirkan Coker and Ares J. Rosakis
Graduate Aeronautical Laboratories, California Institute of Technology
Pasadena, CA, USA 91125

Yonggang Y. Huang
Department of Mechanical and Industrial Engineering
University of Illinois at Urbana-Champaign, Urbana, IL, USA 61801

SUMMARY

Some recent experimental observations of highly dynamic crack growth events in thick unidirectional graphite fiber-reinforced epoxy matrix composite plates are presented. The composite plates were symmetrically (mode-I) and asymmetrically (mode-II) loaded in a one-point bend configuration with an edge pre-notch machined in the fiber direction. The lateral shearing interferometric technique of coherent gradient sensing (CGS) was used in conjunction with high-speed photography. Symmetric, mode-I cracks initiated at 1300 m/s and subsequently accelerated up to the Rayleigh wave speed but never exceeded it. For asymmetric, Mode-II types of loading, the results reveal highly unstable and intersonic, shear-dominated crack growth along the fibers. The intersonic cracks propagated with unprecedented speeds reaching 7400 m/s, more than three times the shear wave speed of the composite, and featured a shock wave structure typical of disturbances travelling with speeds higher than one of the characteristic wave speeds in the solid.

INTRODUCTION

Dynamic crack growth along weak planes is a significant mode of failure in composites and other layered materials. In the past years bimaterial fracture specimens, fabricated by the bonding of a stiff material to a compliant material (featuring a mismatch in wave speeds), have been used to demonstrate the importance of highly transient and dynamic crack growth in heterogeneous solids. It was observed that interfacial crack tip speeds rapidly approached and exceeded the shear wave speed of the more compliant material [1-5]. Thus they reached intersonic speeds with respect to the more compliant material in which intersonic crack tip speed is defined as the speed in the open interval between the shear wave speed and the longitudinal wave speed. For crack tip speeds above the shear wave speed a ray emanating from the crack tip representing a line of strong discontinuity (shear shock wave) was observed experimentally and also predicted theoretically [2]. These high crack tip speeds were obtained under loading conditions that promoted locally shear dominated deformations at the crack tip which were further enhanced by the stress wave mismatch due to the bimaterial nature of the solid [6].

In homogeneous materials and in the absence of a weak prescribed crack growth path, Mode-II crack propagation is not possible because the crack naturally kinks and propagates in a direction that causes a local Mode-I stress state around the crack tip. This is not necessarily true in solids that may be homogeneous regarding their properties but may involve preferable crack paths in the form of weak planes of lower toughness. Unidirectional composites may indeed fit this description if viewed from the "macroscopic" point of view of a homogenized anisotropic theory. Although from the microscopic viewpoint, unidirectional composite materials are locally inhomogeneous and are related to bimaterials, in the sense that they also involve a preferred crack growth direction, and are also composed of a stiffer material (the fiber) bonded together with a more compliant material (the matrix), from a macroscopic point of view such solids can be viewed as homogeneous anisotropic materials as far as their elastic properties are concerned. However, they are still inhomogeneous regarding their fracture toughness properties. Indeed, from both the macroscopic or microscopic view points the common characteristic between unidirectional composites and bimaterials which also seem to be the most relevant is the existence of a weak straight line crack path which may accommodate growing cracks of both modes and not the existence of the two phases. Because of the above observations and because of our previous discovery of intersonic crack growth in bimaterials, unidirectional composites seem to be natural candidates for studying maximum allowable crack tip speeds of cracks of both modes in a system that is macroscopically at least homogeneous.

While the theory excludes intersonic growth of mode-I cracks, it has not excluded the possibility of intersonic mode-II dominated crack growth in isotropic or anisotropic homogeneous elastic solids in the case of a self similar crack growing in the direction of the plane of the crack [7]. Several researchers [7-11] calculated the critical speed at which intersonic crack growth is stable in isotropic materials to be $\sqrt{2}\, c_s$. For stable intersonic crack growth this is the only speed at which the energy release rate is finite. Analysis of the intersonic crack growth in anisotropic materials was developed by Piva and Hasan [12], Huang et al. [8] and Broberg [13]. They predicted that, just as in isotropic materials, intersonic shear crack growth is possible at a specific critical speed, v_c, above the shear wave speed. However, in the case of anisotropic materials, this speed is the product of a complex function of the material properties and the shear wave speed.

To our knowledge, very few experimental studies of dynamic crack propagation in fiber-reinforced composites have been performed. Liu *et al.* [14], investigated quasi-static fracture of polymer composites using the optical technique of CGS. Khanna and Shukla [15] extended the theoretical results of Piva and Viola [16] to in-plane stress and used it to analyze the results of strain gages and to thus determine mode-I stress intensity factors for cracks propagating in unidirectional glass-epoxy composite laminates at constant speed. Lambros and Rosakis [17], using CGS shearing interferometry technique in conjunction with high speed photography, looked at the initiation and growth in thick unidirectional graphite-epoxy composite plates under symmetric 1-pt impact loading. They observed crack tip speeds for mode-I loading which approached 900 m/s or $0.58c_R$, where c_R is the speed of the Rayleigh waves travelling in the direction of the fibers. In another investigation, Lambros and Rosakis [18] also studied the dynamic delamination propagation in real time of a cross-plied laminate subjected to out-of-plane impact.

In this paper we present experimental results of dynamic crack propagation in unidirectional graphite-epoxy composite plates under mode-I and mode-II loading. The optical technique of CGS was used in conjunction with high-speed camera to obtain interferograms during crack growth. Crack tip speeds are then computed from the interferograms and the limiting crack tip speeds are determined for both the mode-I and the mode-II cases.

MATERIAL

A cross-section of the composite material is shown in Fig. 1. The unidirectional graphite/epoxy composite plates were manufactured from 48 layers of graphite fiber and epoxy matrix pre-pregs laid up in the thickness direction to form a 6.3-mm thick plate. The fiber volume fraction in the prepreg is 65% while the fiber diameter is 7.3 μm. The surface on one side of the composite plate was made optically flat by adding a thin layer of epoxy on one surface and then curing the composite specimen upon an optically flat glass plate. This surface was then made specularly reflective by coating with a thin layer of aluminum of 1-2 μm thickness in a vacuum deposition chamber.

The specimen used was a single edge notch specimen of 127 mm width and 203 mm (Fig. 2). A notch of 25 mm length and 1.5 mm width parallel to the fibers was machined using a diamond-saw with the fibers parallel to the shorter edge of the specimen. During the experiments, crack initiated from the blunt notch. The blunted initial

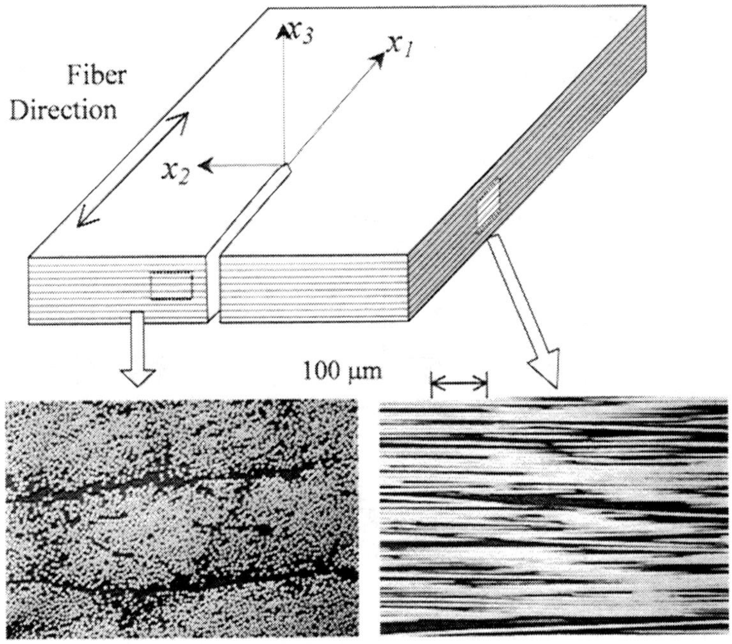

Fig. 1: Cross-section of graphite/epoxy unidirectional composite plate.

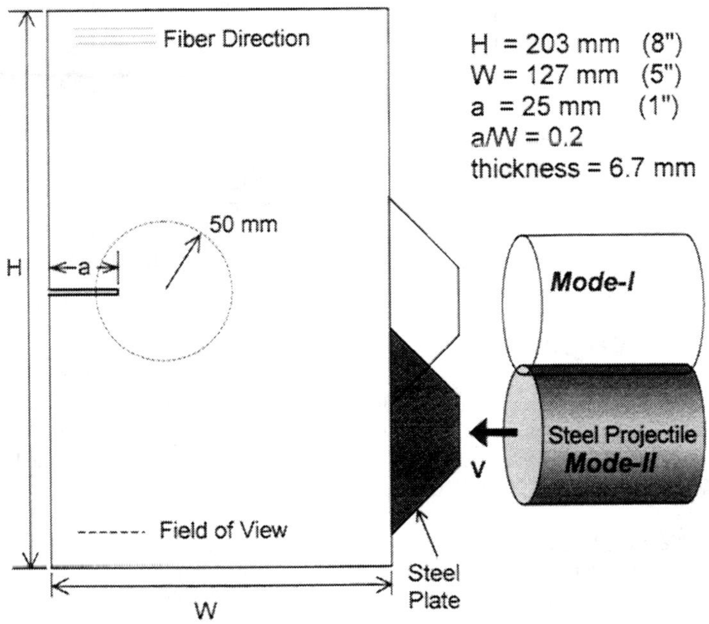

Fig. 2: Single edge notch specimen geometry and dimensions.

notch allows more strain energy to be stored prior to the generation of the dynamically growing crack and thus its bluntness controls (to a certain degree) the initiation crack tip speed.

The orientation of the axes with respect to the composite plate is shown in Fig. 1. The x_1-axis is defined to lie along the fibers, the x_3-axes is perpendicular to the plane of the composite surface while the x_2-axis is perpendicular to the x_1-x_3 plane. The orthotropic elastic constants, the compliance matrix and stiffness matrix values for graphite/epoxy unidirectional composite material are given in Table 1.

Table 1: In-plane material properties of Graphite/Epoxy unidirectional composite.

Elastic Constants		Reduced Stiffness Matrix (GPa)	
E_1	80.0 GPa	c'_{11}	80.5
E_2	8.9 GPa	c'_{22}	9.0
ν_{12}	0.25	c'_{12}	2.2
μ_{12}	3.6 GPa	c'_{66}	3.6

The longitudinal and shear wave speeds in terms of reduced stiffness matrix (plane stress) in the direction of the major axes are given by: $c_l^{||} = (c'_{11}/\rho)^{1/2}, c_l^{\perp} = (c'_{22}/\rho)^{1/2}$, and $c_s = (c'_{66}/\rho)^{1/2}$, in which $c_l^{||}$ denotes the dilatational wave speed parallel to the fibers while $c_l^{\perp}$ denotes the dilatational wave speed perpendicular to the fibers. The density is 1478 kg/m^3. The shear wave speed as well as the longitudinal wave speeds parallel and perpendicular to the fibers are shown in Table 2. The Rayleigh wave speed parallel to the fibers is 1548 m/s

Table 2: Characteristic plane-stress wave speeds.

Wave Speeds	5 MHz Ultrasonic transducer (Plane strain) m/s	(Plane Stress) m/s
$c_l^{\vert\vert} = \sqrt{c'_{11}/\rho}$	7450 ± 100	7380
$c_l^{\perp} = \sqrt{c'_{22}/\rho}$	2740 ± 100	2470
$c_s = \sqrt{c'_{66}/\rho}$	1560 ± 100	1560

EXPERIMENTAL TECHNIQUE

Schematic of the experimental set-up and the optical technique of coherent gradient sensing, CGS, is shown in Fig. 3. The specimen was subjected to impact loading through a projectile fired from a gas gun. The projectile was 75 mm long, 50 mm in diameter and was made of hardened steel. A steel piece was bonded to the specimen at the impact site to prevent crushing of the composite during impact and to induce a planar loading wave front. The projectile speed was varied between 20 to 30 m/s.

The dynamic stress field produced by the impact loading wave leads to a changing out-of-plane deformation on the surface of the composite plate. The optical technique of reflection CGS, in conjunction with high-speed photography, was used to record in real time the slopes of the out-of-plane deformation field. CGS is a full-field, lateral-shearing interferometer. The details of CGS can be found in several articles [19, 20] and its application to crack initiation in composite materials has been demonstrated before [14, 17]. CGS, when used in reflective mode, measures the in-plane gradients of out-of-plane displacements;

$$\frac{\partial u_3}{\partial x_1} = \frac{mp}{2\Delta}, \quad m = 0, \pm 1, \pm 2, ... \tag{2}$$

where $u_3(x_1,x_2)$ is the out-of-plane displacement of the reflective surface of the specimen, p and Δ are the pitch and separation of the two high-density gratings, and m is the fringe order for the x_1 gradient contours. According to the above equation, each CGS fringe is a locus of points that have the same slope in the x_1-direction.

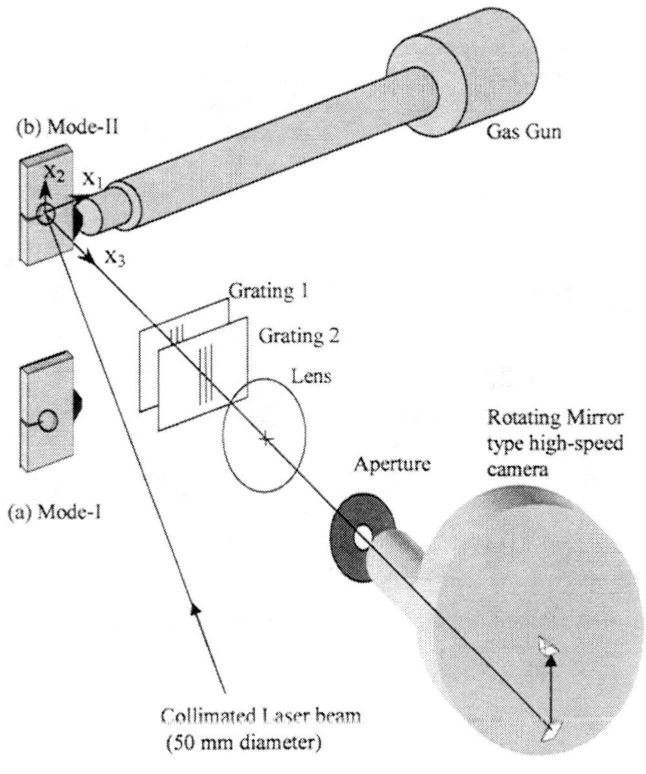

Fig. 3: Schematic of the experimental set-up and optical technique of CGS.

The experimental set-up used a coherent, monochromatic, collimated laser beam of 50 mm diameter which was incident on the specimen and was reflected off the optically flat surface. Subsequently the reflected beam acquired an optical path difference due to the non-uniform specimen surface deformation near the crack tip after impact. The resulting beam passed through two high-density line diffraction gratings G_1 and G_2 which sheared the reflected beam and recombined it with itself to form an interferogram. The interference pattern was finally captured by a rotating mirror type high-speed camera (Cordin model 330A). The camera is capable of recording 80 frames at framing rates up to 2 million frames per second.

EXPERIMENTAL OBSERVATIONS

Symmetric Mode-I crack tip deformations were attained by impacting the specimen symmetrically along the notch line by a steel projectile as shown in Fig. 3 with projectile speeds varying from 10 m/s to 57 m/s. A typical sequence of experimental CGS interferograms for mode-I crack initiation and propagation are shown in Fig. 4 for the highest impact speed of 57 m/s. The first frame shows the CGS interference fringe pattern around the notch tip just before crack initiation. The fringe pattern around the crack tip during propagation at times 1.8 μs and 3.0 μs is shown in Figs. 4(b) and (c). Using the above sequence of pictures the crack tip history was recorded as a function of time in Fig. 5(a) for Mode-I symmetric loading. The crack tip speed history was then obtained by using 3-point fit to the crack tip history and is shown in Fig. 5(b). The crack started growing subsonically at 1350 m/s and accelerated to the Rayleigh wave speed of the composite in the direction of the fibers, 1550 m/s. Regarding small oscillations in the calculated crack tip speeds as experimental error, the crack tip speeds did not exceed the Rayleigh wave speed.

Asymmetric 1-pt bend impact experiments were conducted at different impact speeds by impacting the specimen below the notch as shown in Fig 3(b). Projectile speed varied from 21 m/s to 30 m/s. For an impact speed of 21 m/s a sequence of three frames, selected from a sequence of 80 frames, taken with the high speed camera at an interframe time of 1.39 μs is shown in Fig. 6. The photographs show the CGS interference fringe patterns in a 50-mm diameter area around the notch tip before crack initiation and the first frame after crack initiation. The notch tip as well as the stress concentration at the notch tip is clearly visible in the first frame. The impact wave (compression) has propagated from one end of the plate to the other, was reflected as a tensile wave below the notch and loaded the notch tip in a predominantly shear mode. A shadow of a marker for scaling appears at the top left side. The vertical line in the center is a streak line and is an artifact of the high-speed camera.

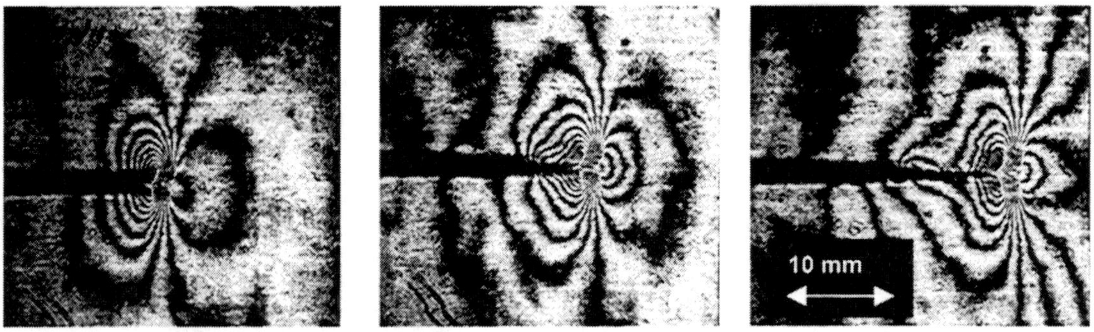

Fig. 4: A close-up view of CGS interferograms for mode-I dynamic crack growth.

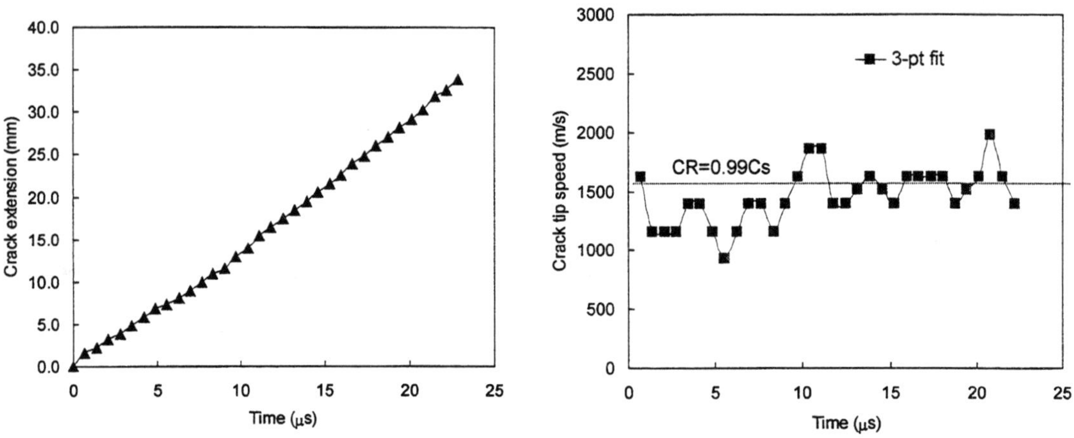

Fig. 5: Crack tip position (a) and crack tip speed (b) history for a crack propagating under symmetric mode-I loading.

Fig. 6: A selected sequence of interferograms up to crack initiation for mode-II loading.

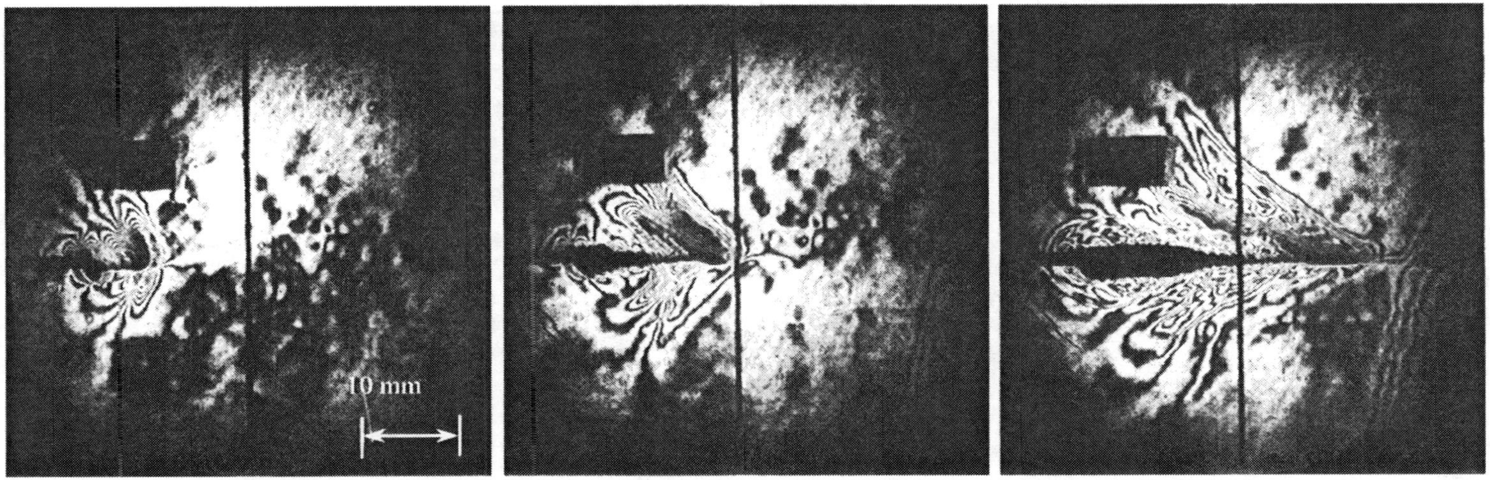

Fig. 7: A selected sequence of CGS interferograms for an intersonically propagating crack tip selected at 2.8 μs time interval. Field of view is around the notch tip.

Notch-tip stress concentration is seen to increase as the interferogram fringes increase in size and number. The nature of the fringe pattern shows that the loading is primarily mode-II [14] featuring one fringe lobe in the back and two fringe lobes in the front which is a distinctive characteristic of asymmetric, mode-II CGS fringe patterns. These pictures show the fringe patterns around the notch tip just before and after crack initiation (last photograph). However, in the last frame, a significant change takes place in which the fringe loops are squeezed backwards indicating very high initial crack tip accelerations. Indeed, the average crack tip speed between these two frames is 2100 m/s, which is above the shear wave speed 1800 m/s for this material. This dramatic change in shape is not observed in the mode-I crack growth experiments described above. The notch tip is loaded by the arrival of the loading wave initially at 11 μs after impact and is fully loaded by the reflected wave at 15 μs. Crack initiates from the notch tip at approximately 32 μs.

Fig. 7 shows a sequence of three CGS interferograms later in time and depicting shear dominated dynamic crack growth. As the crack propagates it accelerates and further dramatic changes can be observed in the shapes of the crack-tip fringe patterns. The fringes are pushed further back and are elongated. The rear loop shape changes from rounded to a triangular wedge bounded by a line of high intensity fringes emerging from the crack tip at an angle. This line is caused by a steep change in the stress gradients in a localized area, which later forms a discontinuity in the shear stresses, i. e. a shear shock wave. Although CGS is sensitive to the sum of the normal stresses in homogeneous, isotropic materials, the normal stresses and shear stresses are coupled in composite materials, a phenomena which allows us to observe discontinuities in shear stresses in our experiments. This would not hold for isotropic solids. Finally, this line broadens into two parallel lines (a double shock wave) and intercepts the crack surface over a finite area between the crack tip and 4-5 mm behind it. We propose for the reason of the double shock wave structure as the existence of a contact region behind the crack tip or alternatively as the cohesive zone in front of the crack tip. The slope of the shear shock wave changes as you move away from the crack tip which may be due to the unsteady growth of the crack tip. Furthermore, as the crack begins to move intersonically, the front fringe loops are also pushed back by yet another line which is not as dramatic as the shock wave described above. Information regarding the crack tip deformation field cannot travel outside this outer line. The normal speed at which this outer line propagates is roughly 2400 m/s which corresponds to the dilatational wave speed of epoxy, one of the two constituents of the composite plate. Thus, this second line of discontinuity seems to represent the dilatational shock wave in the epoxy layer due to a moving source (the crack tip) disturbance. The existence of this line would not be predicted by theoretical derivations based on homogeneous anisotropic theory.

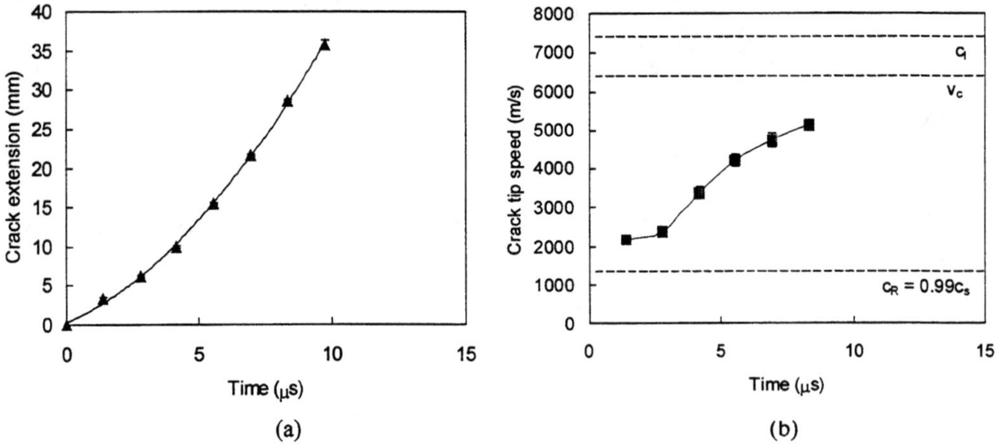

Fig. 8: Crack tip location (a) and speed (b) history for shear-dominated crack growth.

The crack tip location and speed history is shown in Fig. 8(a) and 8(b). The crack-tip speed approaches 5200 m/s, a speed 3 times higher than the shear wave speed and is clearly intersonic. In this experiment the crack tip jumps immediately from rest to 2100 m/s becoming intersonic in the first frame after initiation, thus not passing through the subsonic regime. Indeed, subsonic crack tip speed was not observed in the time resolution of this experiment. The crack tip acceleration was also very high and was of the order of 10^9 m/s^2.

From Fig. 8 it can be seen that the crack propagation had not reached steady state within our window of observation. A separate experiment was conducted under similar conditions with the field of view further downstream of the notch in order investigate whether the crack tip speed eventually attains steady state conditions following the initial acceleration stage. In this case the impact velocity was 28 m/s and the interframe time was 0.83 μs. Fig. 9 shows a selected set of CGS fringe patterns. The crack tip position and speed history are plotted in Fig. 10(a) and (b) respectively. As already observed from the previous case, a shear shock wave and an outer shock wave corresponding to the epoxy is observed. However, in contrast to the previous experiment, a single dark line radiates from the crack tip. The first recorded crack tip speed is 4000 m/s as the crack appears in our field of view. The crack tip speed then climbs up and oscillates around and an average speed of 7000 m/s. This is the highest crack tip speed ever observed in laboratory setting. However, it is inconclusive whether steady state has been reached or not as the slope of the shock wave away from the crack tip is still varying.

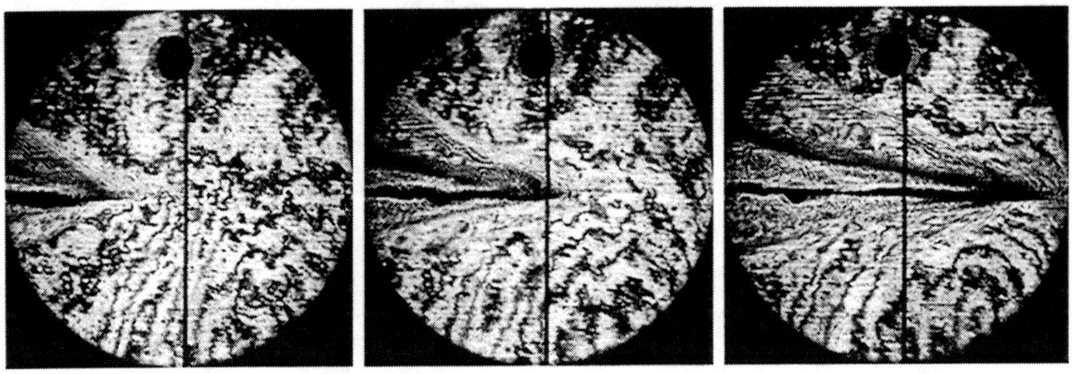

Fig. 9: A selected sequence of CGS interferograms for an intersonically growing shear crack at 1.67 μs time intervals. Field of view ahead of the notch tip.

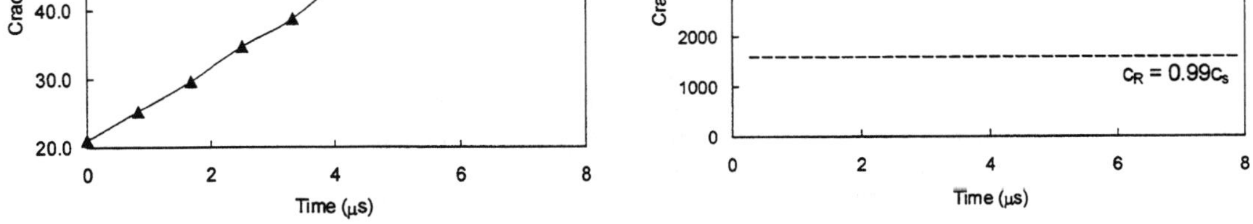

(a) (b)

Fig. 10: Crack tip position (a) and crack tip speed (b) for shear dominated crack growth for a field of view ahead of notch- tip.

DISCUSSION

A view of the general trends in crack tip position history characteristics of many different experiments corresponding to very similar impact conditions can be observed in Fig. 11(a). This figure shows the reproducibility of the observed experimental trends (within the same geometry) and allows us to obtain an average sense of the trends in the variation of the intersonic crack tip speed for both mode-I and mode-II cases. Crack tip speed as a function of crack extension is shown in Fig. 11(b). The crack tip speeds were calculated in two ways: cubic least squares fit to mode-II and mode-I curves (solid lines) and 3-point fit for individual experiments (symbols). Mode-I cracks initiate around 1300 m/s and subsequently the crack speed increases to the the neighborhood of the Rayleigh wave speed of 1548 m/s. However, mode-II cracks inititiate around 2000 m/s (intersonic) and rapidly accelerate above 6000 m/s. These cracks asymptotically approach a steady state speed of 6500 m/s.

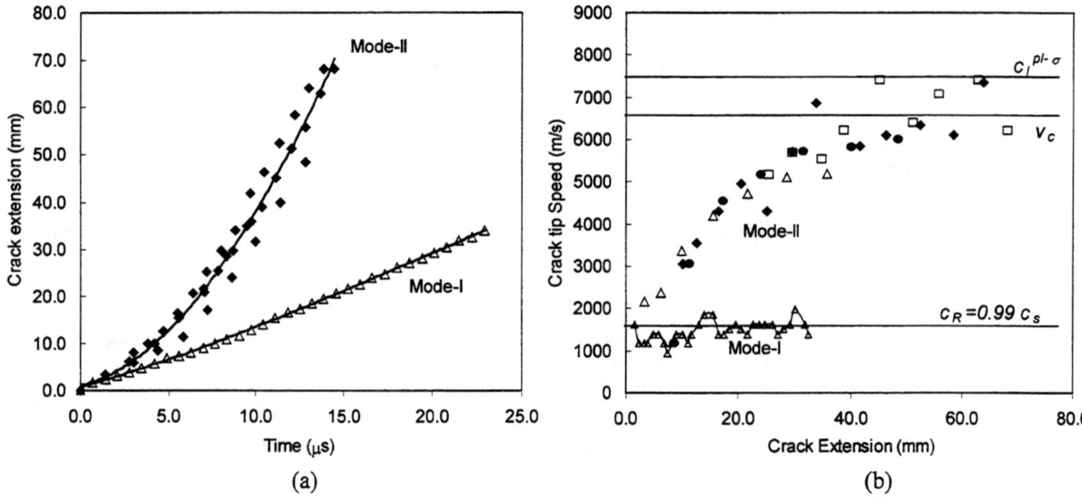

Fig. 11: Crack tip position history (a), and crack tip speed as a function of crack position (b) for both mode-I and mode-II cracks.

In order to interpret the observed phenomena theoretical analysis was developed by Huang et al. [8]. They obtained the asymptotic stress and displacement fields near an intersonically propagating crack tip in an orthotropic material under steady-state conditions. In this analysis, the composite was modeled as an elastic, orthotropic, homogeneous solid. For both mode-I and mode-II cases a prescribed straight-line crack path was assumed. Mode-I elastic asymptotic field yielded a negative, unbounded crack tip energy release rate in the entire intersonic crack growth regime. This is physically impossible since a propagating crack-tip cannot radiate out energy. Thus, a mode-I crack tip cannot propagate intersonically. This conclusion is supported by our experiments in which the crack tip speed never exceeded the shear wave speed, regardless of how large the impact velocity of the projectile was (see Fig. 11b). For mode-II, the energy release rate supplied by the elastic asymptotic field is finite and positive only at a distinct critical crack tip speed [8]. Since a positive and finite energy supply is required to break the material bonds in front of the crack tip this speed corresponds to the only possible steady state intersonic crack growth speed according to the above steady state theory. Using the material properties given in Table 1, this critical crack tip speed is 6600 m/s. In our experiments the mode-II crack tip seem to eventually reach this value sometimes from below and sometimes from above [21].

CONCLUSIONS

Dynamic fracture experiments conducted in real time on prenotched unidirectional fiber reinforced graphite/epoxy composites indicate that in Mode-I (opening cracks) may only propagate subsonically along the fibers with respect to the homogenized anisotropic material properties. Mode-II (shear) cracks are, however, found to be capable of intersonic crack growth along the fibers, typically accelerating from the vicinity of the shear wave speed to the dilatational wave speed along the fiber direction and eventually reaching a steady state at a particular critical intersonic speed, v_c, determined by the homogenized anisotropic material properties. Dual shock waves, which might imply a region of contact, are formed at speeds lower than v_c while a single shock wave becomes visible at higher speeds. In absolute terms, unprecedented maximum crack tip speeds as high as 7,400 m/s are observed.

ACKNOWLEDGMENT

This investigation was supported by the Office of Naval Research (Dr. Y. D. S. Rajapakse, Scientific Officer) through a grant to Caltech and is gratefully acknowledged.

REFERENCES

1. Lambros, J. and A. J. Rosakis, "Shear dominated transonic interfacial crack-growth in a bimaterial: 1. Experimental observations," *Journal of the Mechanics and Physics of Solids*, **43**(2), 1995, 169-188.
2. Liu, C., Y. Huang, Rosakis, A. J., "Shear dominated transonic interfacial crack growth in a bimaterial: 2. Asymptotic fields and favorable velocity regimes," *Journal of the Mechanics and Physics of Solids*, **43**(2), 1995, 189-206.
3. Singh, R. P., J. Lambros, Shukla, A. and Rosakis, A. J., "Investigation of the mechanics of intersonic crack propagation along a bimaterial interface using coherent gradient sensing and photoelasticity," *Proceedings of the Royal Society of London Series A,* **453**(1967), 1997, 2649-2667.
4. Singh, R. P. and Shukla, A., "Subsonic and intersonic crack growth along a bimaterial surface," Journal of Applied Mechanics **63**, 1996, 919-924.
5. Rosakis, A. J., Samudrala, O., Singh, R. P. and Shukla, A., "Intersonic crack propagation in bimaterial systems." *Journal of the Mechanics and Physics of Solids,* **46**(10), 1998, 1789-1813.
6. Lambros, J. and Rosakis, A. J., "Development of a dynamic decohesion criterion for subsonic fracture of the interface between two dissimilar materials," *Proceedings of the Royal Society of London Series A*, **451**(1943), 1995, 711-736.
7. Broberg, K. B., "The Near-Tip Field at High Crack Velocities." *International Journal of Fracture*, **39**, 1989, 1-13.
8. Huang, Y., Wang, W., Liu, C. and Rosakis, A. J., "Analysis of intersonic crack growth in unidirectional fiber-reinforced composites," to be published in *Journal of the Mechanics and Physics of Solids*, 1999.
9. Burridge, R., "Admissible speeds for plane-strain self-similar shear cracks with friction but lacking cohesion," *Geophysical Journal of the Royal Astronomical Society,* **35**, 1973, 439-455.
10. Freund, L. B., "The Mechanics of Dynamic Shear Crack Propagation," *J. of Geophysical Research,* **84**(B5), 1979, 2199-2209.
11. Georgiadis, H. G., "On the stress singularity in steady-state transonic shear crack propagation," *International Journal of Fracture*, **30**, 1986, 175-180.
12. Piva, A. and Hasan, W., "Effect of orthotropy on the intersonic shear crack propagation," *Journal of Applied Mechanics,* **63**, 1996, 933-938.
13. Broberg, K. B., "Intersonic crack propagation in an orthotropic material," *International Journal of Fracture,* to be published, 1999.
14. Liu, C., Rosakis, A. J., Ellis, R. W. and Stout, M. G., "A study of the fracture behavior of unidirectional fiber-reinforced composites using coherent gradient sensing (CGS) interferometry," *International Journal of Fracture,* **90**, 1998, 355-382.
15. Khanna, S. K. and Shukla, A., "Development of stress-field equations and determination of Stress intensity factor during dynamic fracture of orthotropic composite materials," *Engineering Fracture Mechanics* **47**(3), 1994, 345-359.

16. Piva, A. and Viola, E., "Crack-propagation in an orthotropic medium," *Engineering Fracture Mechanics*, **29**(5), 1988, 535-548.
17. Lambros, J. and Rosakis, A. J., "Dynamic crack initiation and growth in thick unidirectional graphite/epoxy plates," *Composites Science and Technology,* **57**(1), 1997, 55-65.
18. Lambros, J. and Rosakis, A. J., "An experimental study of dynamic delamination of thick fiber reinforced polymeric matrix composites," *Experimental Mechanics*, **37**(3), 1997, 360-366.
19. Tippur, H. V., Krishnaswamy, S. and Rosakis, A. J., "A coherent gradient sensor for crack tip deformation measurements: analysis and experimental results," *International Journal of Fracture*, **48**, 1991, 193-204.
20. Rosakis, A. J., "Two optical techniques sensitive to the gradients of optical path difference: The method of caustics and the coherent gradient sensor (CGS)," *Experimental Techniques in Fracture*, J. S. Epstein, Ed., New York, VCH, 1993, 327-425.
21. Coker, D. and Rosakis, A. J., "Experimental Observations of Intersonic Crack Growth in Asymmetrically Loaded Unidirectional Composite Plates," GALCIT SM Report No. 98-16, Pasadena, California Institute of Technology, 1998.

AMD-Vol. 235, Thick Composites for Load Bearing Structures
ASME 1999

DYNAMIC BEHAVIOR OF FIBER REINFORCED COMPOSITES UNDER MULTIAXIAL COMPRESSION

Kenji Oguni and G. Ravichandran
Graduate Aeronautical Laboratories
California Institute of Technology
Pasadena, California

ABSTRACT

Results from an experimental investigation on the mechanical behavior of a unidirectional reinforced polymer composite with 50% volume fraction E-glass/vinylester under uniaxial and proportional multiaxial compression are presented. Specimens are loaded in the fiber direction using a servo-hydraulic material testing system for low strain rates and a Kolsky (split Hopkinson) pressure bar for high strain rates, up to 3000 s^{-1}. The results indicate that the compressive strength of the composite increases with increasing levels of confinement and increasing strain rates. Post-test optical and scanning electron microscopy is used to identify the failure modes. The failure mode that is observed in unconfined specimen is axial splitting followed by fiber kink band formation. At high levels of confinement, the failure mode transitions from axial splitting to kink band formation and fiber failure. Also, a new energy based analytic model for studying axial splitting phenomenon in unidirectional fiber-reinforced composites is presented.

INTRODUCTION

Deformation and fracture behavior of fiber reinforced composites have received considerable attention because of their importance in structural applications. Composites are also used in impact-related applications such as marine structures, turbine blades, automotive and others. Of particular interest for composite structures subjected to impact are their high-strain-rate mechanical properties, resistance to crack initiation and propagation as well as their strength and corresponding failure mechanisms under dynamic loading conditions. Therefore, investigation on the dynamic deformation and fracture behavior of fiber reinforced composites is needed in order to develop a mechanistic understanding of their behavior. Also, in general, the loading in most applications is multiaxial, and even under uniaxial loading, the stress state in a composite laminate is multidimensional due to inter-laminar shear coupling. Hence, understanding of deformation and fracture behavior of fiber reinforced composites is also important. Such an understanding will aid in the development of constitutive and failure models for analysis and design of composite structures subjected to impact (Lambros and Rosakis, 1997 a, b). Relatively little is known concerning dynamic behavior of fiber reinforced composites (Lankford, 1997, Waas et al., 1997).

The limiting factor in the design of composite structures is their compressive strength. The compressive strength of unidirectionally reinforced composites is found to be roughly one-half of their tensile strength. Also, their compressive strength has been consistently and considerably lower than theoretical predictions. Extensive studies have been carried out on unidirectional fiber composites under uniaxial compression (Waas and

Schultheisz, 1996). In general, the loading in most applications such as pressure vessels and submersibles is multiaxial. Even under uniaxial loading, due to shear coupling the stress state in a laminate is multi-dimensional. However, little is known concerning the multiaxial behavior of fiber reinforced composites (Lankford, 1997, Weaver and Williams, 1975, Parry and Wronski, 1982). The limited work concerning behavior of composites under multiaxial compression has been performed under hydrostatic pressure. In many applications involving composites, e.g. laminates, the loading path is proportional, i.e. stress components change in proportion to one another.

A major objective of this paper is to present a new experimental technique for studying the dynamic behavior of fiber reinforced composites using a modified Kolsky (split Hopkinson) pressure bar. This technique is used to study the high-strain-rate behavior of a unidirectionally reinforced E-glass/vinyl ester polymeric composite under multiaxial compression. The deformation and failure response of the composite over a wide range of strain rates and stress states are presented and discussed. A new energy-based analytic model for studying axial splitting phenomenon in unidirectionally fiber-reinforced composites is presented.

EXPERIMENTAL

Modified Kolsky Pressure Bar

Kolsky (Split Hopkinson) pressure bar is a well-established apparatus commonly utilized in the high-strain-rate testing of materials. Originally developed by Kolsky (1949), the concept has found widespread applications in testing ductile materials at strain rates up to 10^4 s^{-1}. However, the application of this technique without adequate modifications for testing composite materials has serious limitations. As will be discussed below, modifications must be made to the conventional split Hopkinson pressure bar to reliably obtain properties at small strains as well as to avoid repeated loading of the specimen thus enabling specimen recovery. The modified Kolsky (Split Hopkinson) bar is shown in Fig. 1.

The conventional Kolsky pressure bar consists of a striker bar, an incident bar and a transmission bar. A specimen made of the material under investigation is placed between the incident bar and the transmission bar. When the striker bar impacts the incident bar, an elastic compressive stress pulse, referred to as the incident pulse, is generated and propagates along the incident bar towards the specimen. The pulse duration equals the round-trip time of a longitudinal elastic bar wave in the striker bar. When the incident pulse reaches the specimen, part of the pulse is reflected back in the incident bar due to impedance mismatch at the bar/specimen interface, and the remaining part is transmitted through the specimen into the transmission bar. The strain gages mounted on the bars provide time-resolved measures of the pulses in the incident and the transmission bars. For a specimen that is under mechanical equilibrium, Kolsky (1949) showed that the nominal strain rate $\dot{\varepsilon}$ (t) in the specimen could be calculated using the relation

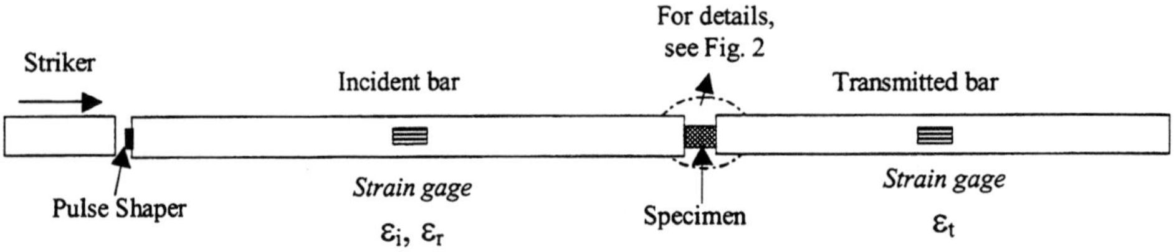

Fig. 1: Schematic of a modified Kolsky (split Hopkinson) pressure bar

$$\dot{\varepsilon}(t) = -\frac{2c_o}{l}\varepsilon_r(t) \tag{1}$$

where l is the original length of the specimen, $\varepsilon_r(t)$ is the time-resolved reflected strain measured in the incident bar, and c_0 $\left(\sqrt{E/\rho}\right)$ is the longitudinal bar wave speed in the bar material. E and ρ are the Young's modulus and the mass density respectively. Integration of Eq. (1) with respect to time gives the time-resolved axial strain of the specimen.

The nominal axial stress σ in the specimen is determined using the equation

$$\sigma(t) = E\frac{A_o}{A_s}\varepsilon_t(t) \tag{2}$$

where A_s is the cross-sectional area of the specimen, and $\varepsilon_t(t)$ is the time-resolved strain in the transmission bar of area A_0. All the foregoing calculations are based on the assumption that the specimen undergoes homogeneous deformation. In the derivation of Eq. (1) and Eq. (2), the incident and transmission bars are assumed to be of the same material, remain elastic and of identical and uniform cross-sectional area.

When nominally brittle materials such as composites are tested in the conventional split Hopkinson pressure bar, the limitations of the technique must be recognized. In order to obtain reliable and consistent experimental data when testing these materials with the Kolsky bar, appropriate modifications must be incorporated in both the experimental technique and the design of specimen geometry. For example, shaping of the loading pulse by a thin soft disc, called a pulse shaper, placed at the impact end of the incident bar has been used to prevent brittle high strength materials from failing before equilibrium is attained in the specimen. In addition to pulse shaping, reliable strain data at small strains (<1%) has been obtained during testing of brittle materials by mounting strain gages on the specimen surface (Chen and Ravichandran, 1997). The limiting strain rate below which reliable deformation and failure data for a brittle material can be obtained using the split Hopkinson pressure bar technique has been established (Ravichandran and Subhash, 1994). The stress in the specimen is computed from the transmitted pulse using Eq. (2) and has been shown to be in close agreement with the stress in the specimen (Chen et al., 1996).

Using the conventional split Hopkinson pressure technique, it is possible for the specimen to be loaded multiple times due to subsequent wave reflections in the incident bar. In the investigation, the transmission bar was made to be shorter than the incident bar (Chen and Ravichandran, 1996) as shown in Fig. 1. With this modification, the shorter transmission bar will act as a momentum trap, thereby moving the transmission bar away from the specimen before a second compressive pulse due to reflected tensile pulse in the incident bar reloading the specimen. Thus, the specimen having been subjected to a single known loading pulse can be recovered for microstructural characterization and unambiguous interpretation of failure modes.

Experimental Setup

The dimensions of the bars in the Kolsky pressure bar setup used in this study are 1220 and 580 mm in length for the incident and transmission bar respectively, with a common diameter of 19 mm. The striker bars are also of 19 mm diameter varied in their lengths from 50 to 100 mm to achieve the desired loading pulse duration. All the bars are made of high strength maraging steel (C-350, Rockwell hardness, Rc=60) with a yield strength of 2.7 GPa. A thin, half-hardened copper disc of 3 mm diameter and 0.85 mm in thickness is typically used as a pulse shaper. The material as well as the diameter and the thickness of the pulse shaper are varied to control the rise time of the incident pulse. The rise time and shape of the pulse are tailored to ensure stress equilibration within the specimen (Ravichandran and Subhash, 1994). High resistance (1000 Ω) strain gages (Micro-measurements WK-06-250BF-10C) with excitation voltage of 30 volts are used to measure the surface strain on the specimen as well as on the bars. Raw strain gage signals without any pre-amplifiers that may distort the signals are recorded using a high-speed 12-bit digital oscilloscope, Nicolet model 440.

Lateral Confinement

A schematic for imposing proportional lateral confinement on a specimen that is axially loaded are shown in Figs. 2 a and b. The experimental set-up consists of a cylindrical specimen placed in a hollow cylinder with a sliding/running fit and is axially loaded using platens. The hollow cylinder and the loading platens are designed to remain elastic during the experiments. The confining cylinder and the platens are made of high strength alloys. Proportional loading is achieved by proper choice of the geometry (the inner and outer radii, a and b respectively) and the material properties for the hollow cylinder. The lateral confinement σ_c, in the elastic regime is a function of the axial stress σ, the cylinder geometry (Fig. 2b) and the elastic properties of the composite specimen and the confining cylinder,

$$\sigma_c = \frac{\sigma}{\left[\frac{(1-\nu)E_{22}a^2}{\nu_{12}E(b^2-a^2)}\left\{1+\frac{(1+\nu)\,b^2}{(1-\nu)\,a^2}\right\}+\frac{(1-\nu_{32})}{\nu_{12}}\right]} \tag{3}$$

where E and ν are the Young's modulus and Poisson's ratio of the sleeve material and E_{22}, ν_{12} and ν_{32} are the elastic properties for the composite.

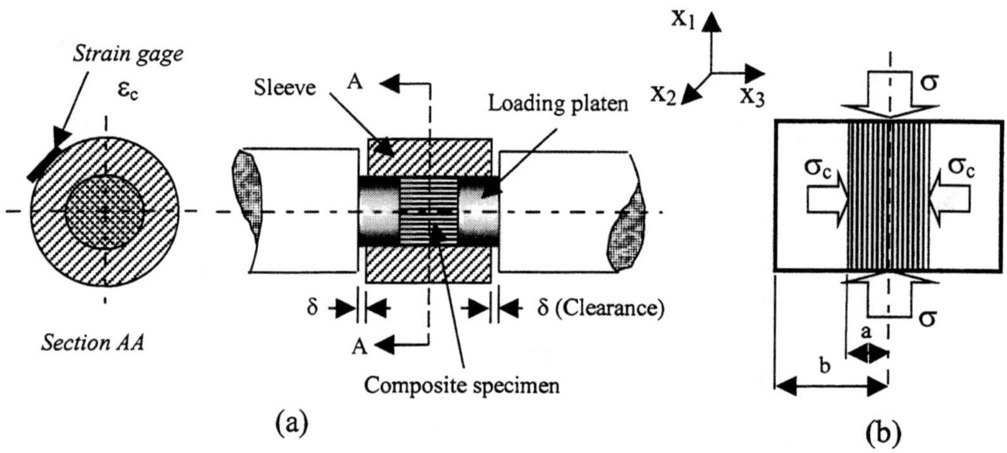

Fig. 2: (a) Schematic of a composite specimen surrounded by a metallic sleeve and loaded through hardened platens; (b) Geometry and stress state of an ideally confined composite specimen

With the guidance from Eq. (3) and proper material choice, desired lateral confinement can be attained using the hollow cylinder configuration in Fig. 2. A strain gage mounted on the external surface of the confining cylinder is used to measure the circumferential or hoop strain (ε_c) and to ascertain the confining stress, $\sigma_c=E\varepsilon_c$. Such a configuration to apply proportional confinement loading can be used both under quasistatic and dynamic loading conditions.

Materials

A unidirectional fiber reinforced composite with 50% volume fraction E-glass/vinyl ester is investigated in the present study. This material is finding increasing applications in marine structures because of the relatively low cost in manufacturing using techniques such as resin transfer molding (RTM). Continuos E-glass (Certainteed R099-625) fibers of 24.1 μm in diameter are aligned in a glass tube and are impregnated with vinyl ester resin (Dow Derakane 411-C50). Following curing, specimens of desired length (6.25 mm) are sectioned using a low speed diamond saw and are sized to desired diameter (6.25 mm) using low speed machining. The ends of the specimen are made parallel and polished using diamond paste. The details of the material and specimen preparation can be found elsewhere (Waas et al., 1997).

The confinement sleeve (Fig. 2a) is typically made of a 7075 aluminum alloy and the dimensions are chosen to provide the desired confinement level (Eq. (3)). The inner diameter is carefully machined to provide smooth sliding fit on the specimen as well as the hardened sleeves. The inner and outer diameters of a typical sleeve used in the experiments are 6.25 mm and 30 mm respectively. The loading platens are made of hard tool steel, Rockwell hardness Rc=60 and dimensions 6.25 mm in diameter and 2.5 mm in length. The lengths of the sleeve and the platens are chosen to provide a predetermined clearance (δ) used to control the extent of deformation imposed on the specimen.

RESULTS

Experiments on the unidirectionally fiber reinforced E-Glass/vinyl ester composite material were performed at low stain rates using a servo hydraulic materials testing system and at high strain rates using the modified Kolsky (Split Hopkinson) pressure bar. Experiments were conducted in the strain rate range of 0.001 to 3,000/s. Experiments were also performed on the pure matrix material, vinyl ester (Dow Derakane 411-C50).

Stress-Strain Response

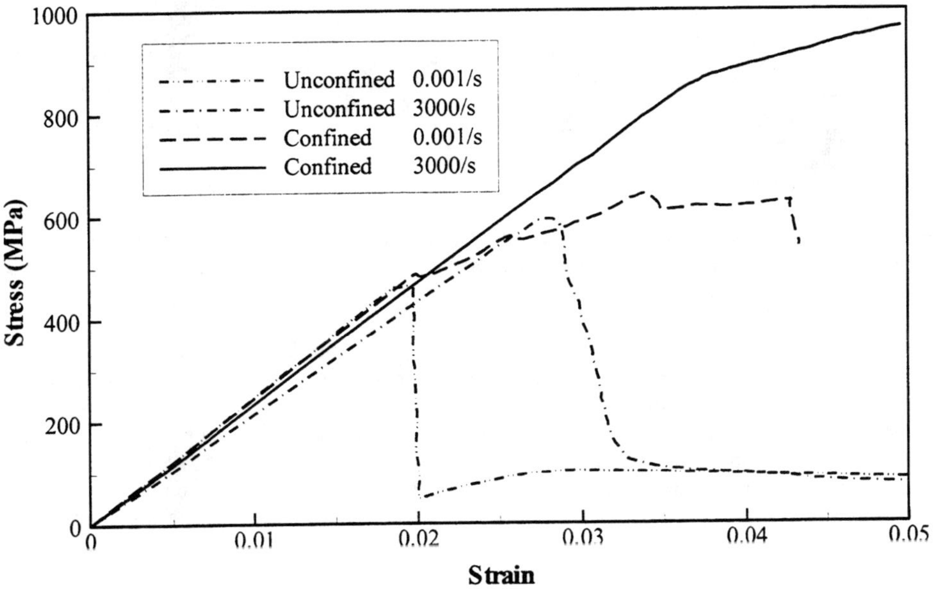

Fig. 3: Stress-strain curves for unconfined and confined 50% V_f E-glass/vinylester composite

The stress-strain curves obtained from experiments for the unconfined composite specimens loaded in the fiber direction are presented here for nominal axial strain rates of 0.001and 3,000/s in Fig. 3. The stress-strain curves are linear up to a maximum prior to catastrophic load drop and are followed by deformation at a nearly constant stress. The peak stress increased from 470 MPa at a strain rate of 0.001/s to 600 MPa at a strain rate of 3,000/s. All the specimens in the above uniaxial experiments failed by axial splitting. The multiaxial compression experiments were designed for the stress ratio σ_c/σ of 0.3 (Eq. (3)). The axial stress-strain curves for the multiaxial compression experiments loaded in the fiber direction and confined laterally (Fig. 2) are shown in Fig. 3 for the strain rates of 0.001 and 3,000/s. The experiments were stopped at a strain of 0.05 to enable the failure mode characterization of the specimens. The maximum stress attained during the experiments on confined specimens increased with increasing strain rate from 600 MPa at 0.001/s to 1,000 MPa at 3,000/s. At low strain rate (0.001/s), the material exhibited a linear response up to 400 MPa followed by load drops and degradation of modulus. Extensive acoustic emission activity was observed during these load drops. At a given strain rate, the maximum stress that the material appears to sustain under multiaxial compression is greater than its unconfined strength. Following the deformation, the confined specimens were sectioned and all the specimens appeared to be intact without any signs of visible failure.

The response of the matrix under unconfined conditions was highly non-linear with yielding at approximately 50 MPa and exhibits a relatively low hardening (Waas et al., 1997). When the matrix was subjected to multiaxial compression under lateral confinement, the matrix exhibited a strongly pressure sensitive behavior. Under these conditions, the matrix did not yield even at 600 MPa and the axial modulus remained unchanged with confinement.

Failure Modes

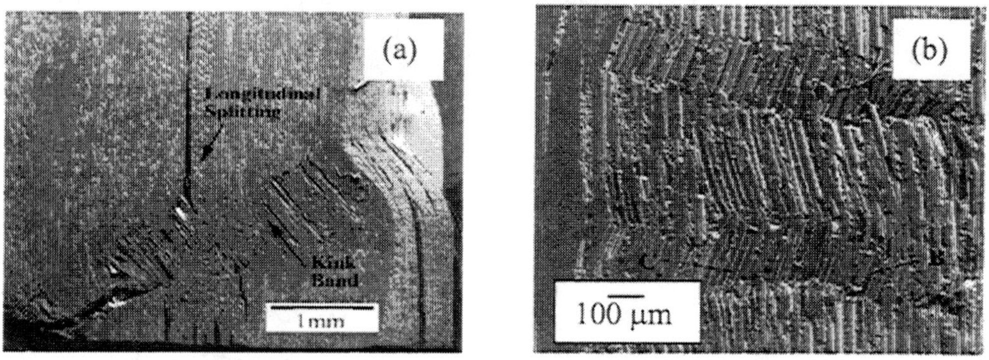

Fig. 4: Failure mode transition in E-glass/vinylester composite, (a) axial splitting followed by kink banding (unconfined); (b) Kink banding followed by fiber fracture (laterally confined)

The failure surfaces of the specimen from the uniaxial experiments and the sectioned surfaces of the specimen from the multiaxial experiments were examined using optical and scanning electron microscopy. The failure surfaces of the unconfined specimens revealed axial splitting induced kink banding as shown in Fig. 4a. The specimen splitting appeared to have preceded by debonding of the fiber leading to local stiffness reduction. The local softening lead to transverse tension causing the specimen to split. The splitting resulted in relaxation of the stress state in the surrounding matrix leading to microbuckling and kink band formation and subsequent fiber failure. The axial splitting is manifested as a catastrophic drop in the stress as seen from the stress-strain response of the unconfined composite. The post failure modes of the kink band formation results in sustaining a lower level of resistance observed in the stress-strain curve as seen in Fig. 3.

In the confined specimens, multiple or conjugate kink bands are observed on the sectioned surface. The kink banding eventually lead to fiber failure at the ends of the kink bands. The acoustic emission activity observed during the quasistatic experiments in confined specimens corresponds to fiber failure from the kink band formation. The multiple load drops observed in the stress-strain curve are related to the transfer of axial loading in the transverse direction due to the kink band geometry. With increasing confinement, the axial splitting mode of failure in the unidirectional fiber reinforced composite is suppressed and transitions to kink band formation is shown in Fig. 4b. Due to lateral confinement, the axial splitting mode is suppressed and the micro buckling leading to kink band is favored. To accommodate applied axial deformation, multiple (conjugate) kink bands are formed which eventually leads to a limiting strength which is related to the crushing of the composite as observed from the leveling of the stress-strain curve for the confined specimens. The observed failure modes and their transition is a function of the strength and toughness of fiber and matrix and their interface, elastic and inelastic properties of the constituents and the applied stress state.

ENERGY-BASED MODEL OF AXIAL SPLITTING

Motivated by the preceding experimental observation, an energy-based model has been constructed to investigate the failure mechanism for low level of lateral confinement, i.e., axial splitting (Oguni and Ravichandran, 1999). Due to the heterogeneity and anisotropy of the fiber reinforced composite, excessive elastic energy is stored in the composite under compression. Axial splitting can be regarded as a process in which the excessive elastic energy is released through the formation of new surfaces. Therefore, when the reduction of the stored elastic energy by splitting compensates the surface energy, the specimen splits. This energy-based failure criterion combined with the effective properties of the composite based on the elastic properties of the matrix and the fiber provides an analytical expression for the critical stress (compressive strength) for axial splitting. This expression illustrates the effect of material properties, surface energy, fiber volume fraction, fiber diameter, and lateral confining pressure on the critical axial compressive stress for axial splitting. The predictions break down for large confining pressures due to failure mode transition to kinking which is not accounted for in the present model.

Problem Formulation

Consider a cylindrical specimen of an ideal unidirectional fiber reinforced composite under lateral confining stress, σ_c, and axial compressive stress, σ, shown schematically in Fig. 5(a). Under this setting, compare two configurations shown in Fig. 5: (a) one is unsplit, and (b) the other is totally split in the fiber direction. Let the total potential energy density of unsplit and split specimen be Π_u and Π_s, respectively. Comparison between Π_u and Π_s provides the critical axial stress for splitting under given lateral confining stress, σ_c. The criterion for longitudinal splitting is the minimization of the total potential energy density of the specimen. In other words, when Π_u exceeds Π_s, the specimen splits (Bhattacharya et al., 1998).

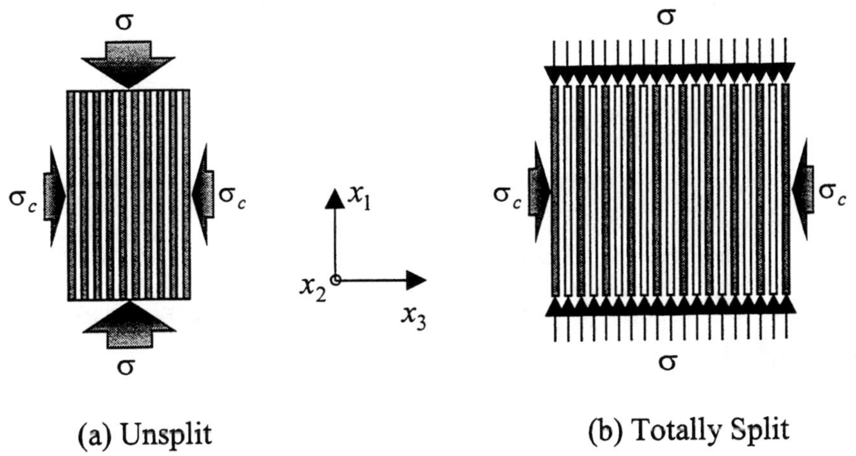

Fig. 5: Schematics of unsplit and axially split configurations for a unidirectional fiber reinforced composite

The total potential energy density of the unsplit specimen, Π_u, is the same as the elastic energy density. Hence, under stress (traction) boundary condition shown in Fig. 5(a), Π_u is expressed as a quadratic form of the normal stress components as follows:

$$\Pi_u = -\frac{1}{2}\left\{\frac{\sigma^2}{E_1^*} + \frac{4\nu_{21}^*\sigma_c\sigma}{E_1^*} + \left(\frac{1}{K_{23}^*} + \frac{4\nu_{21}^{*\,2}}{E_1^*}\right)\sigma_c^{\,2}\right\}. \tag{4}$$

where, independent effective moduli, namely, the longitudinal Young's modulus, E_1^*, Poisson's ratio, ν_{21}^*, the plane strain bulk modulus, K_{23}^* and the shear modulus, G_{23}^* are used. Expressions for these effective moduli have been obtained by Hashin and Rosen (1964).

Under the same boundary condition as that of the unsplit specimen and assuming that *each* RVE (Representative Volume Element) splits at the boundary of the matrix and the fiber, i.e., the split is caused by an interfacial crack (delamination), the split RVE can be regarded as two columns, consisting of either the fiber or the matrix. Such a simplifying assumption enables gaining insights into the strength of composites. The elastic energy density of the RVE after splitting, E_s, is given as a quadratic form of the normal stress components as follows:

$$E_s = -\frac{1}{2}\left[\frac{v_f}{E_f}\left\{\sigma^2 - 4\nu_f\sigma\sigma_c + 2(1-\nu_f)\sigma_c^{\,2}\right\} + \frac{(1-v_f)}{E_m}\left\{\sigma^2 - 4\nu_m\sigma\sigma_c + 2(1-\nu_m)\sigma_c^{\,2}\right\}\right]. \tag{5}$$

where, (v_f, v_m) are volume fractions, (E_f, E_m) are Young's moduli and (ν_f, ν_m) are Poisson's ratios of fiber and matrix, respectively.

The surface energy per unit volume, Γ, of the RVE due to splitting can be obtained by introducing a surface energy per unit area, γ

$$\Gamma = \frac{2\gamma A}{\Omega} = \frac{2\gamma\,(2\pi\,a h)}{\pi R^2 h} = \frac{4\gamma\, v_f}{a} \tag{6}$$

where, A is the lateral surface area of a fiber in the RVE, a is the radius of the fiber and R is the radius of the RVE.

The total potential energy density of the split specimen, Π_s, is the sum of the elastic energy density, E_s and the surface energy density, Γ

$$\Pi_s = E_s + \Gamma . \tag{7}$$

Criterion for Axial Splitting

From the principle of minimum potential energy, the criterion for axial splitting can be expressed as

$$\begin{aligned} \Pi_u - \Pi_s < 0 &\;\Rightarrow\; \text{Unsplit} \\ \Pi_u - \Pi_s = 0 &\;\Rightarrow\; \text{Neutral} \\ \Pi_u - \Pi_s > 0 &\;\Rightarrow\; \text{Split} \end{aligned} \tag{8 a, b, c}$$

The critical condition is given by the equality,

$$\Pi_u - \Pi_s = 0 . \tag{9}$$

Substituting for Π_u and Π_s from Eq. (4) to Eq. (7), the critical stress for longitudinal splitting can be obtained by solving Eq. (9). Since the form of the total potential energy is a quadratic of σ, there are two roots σ_1 and σ_2

$$\sigma_{1,2} = \frac{p_2\sigma_c \pm \sqrt{p_2^2\sigma_c^2 - p_1(p_3\sigma_c^2 - \Gamma)}}{p_1} \tag{10}$$

where p_1, p_2 and p_3 are expressed in terms of the elastic constants of the mterials

$$p_1 = \frac{1}{2}\left(\frac{v_f}{E_f} + \frac{v_m}{E_m} - \frac{1}{E_1^*}\right), \quad p_2 = \frac{v_f \nu_f}{E_f} + \frac{v_m \nu_m}{E_m} - \frac{\nu_{21}^*}{E_1^*},$$

$$p_3 = \frac{v_f(1-\nu_f)}{E_f} + \frac{v_m(1-\nu_m)}{E_m} - \frac{1}{2}\left(\frac{1}{K_{23}^*} + \frac{4\nu_{21}^{*2}}{E_1^*}\right).$$

For a given confining pressure σ_c and surface energy density γ, $\sigma_1 \geq \sigma_2$, hence, σ_1 is taken as the critical stress, σ^*.

Examining the quadratic form of the energy surface, $\Phi(\sigma, \sigma_c) = \Pi_u - \Pi_s$, and assumimg that the longitudinal (fiber direction) compliance is smaller than the lateral (transverse) compliance in the composite (typical for most fiber reinforced composites) the following inequality holds:

$$\frac{d\sigma^*}{d\sigma_c} \leq 1 \tag{11}$$

subject to the constraints $\sigma > \sigma_c$ and $d\Phi = \frac{\partial\Phi}{\partial\sigma}d\sigma + \frac{\partial\Phi}{\partial\sigma_c}d\sigma_c = 0$

From Eq. (11), one can conclude that if the splitting failure is governed by the principle of minimum total potential energy, the slope of the relationship between compressive strength and confining pressure, i.e., σ^* vs. σ_c, can not exceed unity. The effect of lateral confinement and material properties on the compressive strength of composites can be investigated by using Eq. (10).

SUMMARY

A new experimental technique using a modified Kolsky (split Hopkinson) pressure bar has been developed to investigate materials at high strain rates and under multiaxial compression. Methods for proportional radial confinement, pulse shaping, specimen recovery and controlling specimen deformation have been outlined. Experiments on a 50% by volume E-glass/vinyl ester composite at various strain rates of up to 3,000/s revealed an increase in compressive strength with increasing strain rate. Results from proportional loading multiaxial compression experiments on the composite showed an increase in strength in comparison to its unconfined compressive strength at a given strain rate. The experimental data is currently being used to develop high-strain rate constitutive models for fiber reinforced composites as a function of stress-state.

An energy-based model for axial splitting has been developed for predicting the compressive strength of unidirectional fiber reinforced composites under weak lateral confinement. For a given lateral confinement, the compressive strength can be computed as a function of the effective properties of the unsplit and the split composite as well as the fracture energy. The results from the analysis indicate that the effect of confinement on compressive strength is relatively small in comparison to the effect of fiber volume fraction (Oguni and Ravichandran, 1999). The rate of increase in compressive strength with increasing confinement is shown less than or equal to unity. The splitting analysis is able to capture the essential features of experimental data (Weaver and Williams, 1975, Parry and Wronski, 1982, Oguni and Ravichandran, 1998) for unidirectional fiber reinforced composites. Insights gained from the modeling regarding the influence of various material parameters and length scales on the strength of composites are useful in designing marine and other structures with composites.

ACKNOWLEDGMENTS

The research reported in this paper is supported by the Office of Naval Research (Dr. Y. D. S. Rajapakse, Scientific Officer) through a grant to the California Institute of Technology and is gratefully acknowledged. We thank Professor A. M. Waas, University of Michigan for providing the composite specimens used in this study.

REFERENCES

Bhattacharya, K., Ortiz, M., and Ravichandran G., 1998, "Energy-based model of compressive splitting in heterogeneous brittle solids," *J. Mech. Phys. Solids*, Vol. 46, pp. 2171-2181.

Chen, W., Subhash, G. and Ravichandran, G., 1994, "Evaluation of Ceramic Specimen Geometries used in a Split Hopkinson Pressure Bar," *Dymat Journal*, Vol. 1, pp. 193-210.

Chen, W. and Ravichandran, G., 1996, "An Experimental Technique for Imposing Dynamic Multiaxial-Compression with Mechanical Confinement," *Experimental Mechanics,* Vol. 36, pp. 435-438.

Chen, W. and Ravichandran, G., 1997, "Dynamic Failure of a Glass Ceramic under Lateral Confinement," *Journal of the Mechanics and Physics of Solids,* Vol. 45, pp. 1303-1328.

Hashin, Z. and Rosen, B. W., 1964, "The elastic moduli of fiber-reinforced materials," *J. Appl. Mech.*, Vol. 31, pp. 223-232.

Kolsky, H., 1949, "An Investigation of the Mechanical Properties of Materials at Very High Rates of Loading," *Proceedings of Royal Society, London,* Vol. B62, pp. 676-700.

Lambros, J. and Rosakis, A. J., 1997a, "Dynamic Crack Initiation and Growth in Thick Unidirectional Graphite/Epoxy Plates," *Composites Science and Technology*, Vol. 57, pp. 55-65.

Lambros, J. and Rosakis, A. J., 1997b, "An Experimental Study of Dynamic Delamination of Thick Fiber-Reinforced Polymeric Matrix Composites, *Experimental Mechanics*, Vol. 37, pp. 360-366.

Lankford, J., 1997, "Shear Versus Dilatational Damage Mechanisms in the Compressive Failure of Fiber-Reinforced Composites," *Composites*, Vol. 28A, pp. 215-222.

Oguni, K. and Ravichandran, G., 1998, "Dynamic Behavior of Fiber Reinforced Composites Under Multiaxial Compression," *Proceedings of the 11th International Conference on Experimental Mechanics*, Oxford, U.K., I. M. Allison, I. M., Ed., pp. 211-216.

Oguni, K. and Ravichandran, G, 1999, "An Energy-Based Model of Longitudinal Splitting In Unidirectional Fiber Reinforced Composites," *SM Technical Report*, Graduate Aeronautical Laboratories, California Institute of Technology, Pasadena, California.

Parry, T. V. and Wronski, A. S., 1982, "Kinking and Compressive Failure in Uniaxially Aligned Carbon Fiber Composite Tested Under Superposed Hydrostatic Pressure," *Journal of Materials Science,* Vol. 17, pp. 893-900.

Ravichandran, G. and Subhash, G., 1994, "Critical Appraisal of Limiting Strain Rates for Testing Ceramics in a Split Hopkinson Pressure Bar," *Journal of the American Ceramic Society*, Vol. 77, pp. 263-67.

Waas, A. M. and Schultheisz, C. R., 1996, "Compressive Failure of Composites, Part 2: Experimental Studies," *Progress in Aerospace Science,* Vol. 32, pp. 43-78.

Waas, A. M., N. Takeda, N., J. Yuan, J., and Lee, S. H., 1997, "Static and Dynamic Compressive Behavior of Glass Fiber Reinforced Unidirectional Composites," *Proceedings of the American Society for Composites, Twelfth Technical Conference*, Dearborn, Michigan, pp. 552-561.

Weaver, C. W. and Williams, J. G., 1975, "Deformation of a Carbon-Epoxy Composite Under Hydrostatic Pressure," *Journal of Materials Science,* Vol. 10, pp. 1323-1333.

AMD-Vol. 235, Thick Composites for Load Bearing Structures
ASME 1999

INFLUENCE OF DELAMINATION ON STRENGTH OF EXTERNALLY PRESSURIZED GLASS/EPOXY CYLINDERS

Peter Davies

Marine Materials Laboratory
IFREMER
29280 Plouzané, France

Leif A. Carlsson

Department of Mechanical Engineering
Florida Atlantic University
Boca Raton, FL 33431, USA

ABSTRACT

The delamination resistance of filament wound glass/epoxy cylinders has been characterized for a range of winding angles and fracture mode ratios using beam fracture specimens. The results reveal that the fracture resistance increases with increasing winding angle and mode II (shear) fraction (G_{II}/G). It was also found that interlaced fiber bundles in the filament wound cylinder wall acted as effective crack arresters in mode I loading. To examine the sensitivity of delamination damage on the implosion behavior of cylinders, external pressure tests were performed on filament-wound glass/epoxy composite cylinders with artificial defects and impact damage. The results revealed that the cylinder strength was insensitive to the presence of single delaminations but impact damage caused reductions in failure pressure. The insensitivity of the failure pressure to a single delamination is attributed to the absence of buckling of the delaminated sublaminates before the cylinder wall collapsed. The impacted cylinders contained multiple delaminations, which caused local reduction in the compressive load capability and reduction in failure pressure.

INTRODUCTION

Glass/epoxy composite cylinders are widely used, designed to resist either internal pressure (e.g. fluid transport and storage, cooling systems) or external pressure (e.g. underwater applications). One of the main concerns with these structures is the propagation of interlaminar defects, produced during manufacture or service. A sound knowledge of the factors affecting delamination resistance of the composites used for cylindrical structures is essential if their safe working envelope is to be defined. The analysis and test methods developed here are necessary for conducting a fracture mechanics analysis of the severity of defects in composite cylinders, and hence provide the basis for a more rational approach to both NDT after manufacture (acceptance/ rejection criteria) and to in-service inspection. Several tests have been developed to characterize the delamination resistance of composite materials, see e.g. a recent publication of Davies et al. [1], but such tests suffer from several drawbacks when cylindrical composite structures are considered. The standard test methods are developed for high-modulus unidirectional composites. Very little work has been reported on other types of composite systems and lay-ups, although cylindrical structures commonly involve low-modulus glass fibers and angle-ply laminates. Furthermore, almost all previous work has considered fracture specimens with rectangular (flat) cross-section, while specimens machined from cylinders have curvature in one direction. Riddle and Beckwith [2] tested mode I and mode II specimens machined into flat coupons from thick carbon/epoxy cylinders. Davies and Rannou [3] performed DCB tests on curved glass/epoxy specimens cut from cylinders and analyzed the data using compliance calibration which neglects several details of the curved laminate specimens, that may influence measured toughness values.

The objectives of the study are to examine the delamination resistance of filament wound composite cylinders and the influence of delamination damage on the failure of externally pressurized composite cylinders. Pressure tests on cylinders without implanted defects and with impact damage and implanted defects to simulate fabrication and in-service damage, were performed, and the relation to measured delamination toughness is discussed.

Materials and Specimens

Delamination specimens were machined from composite cylinders consisting of E-glass fibers in an epoxy resin. The internal diameter of the cylinders was 160 mm and the nominal wall thicknesses were 6 and 12 mm (12 and 24 layers). The lay-ups of the cylinders were $[\pm\theta]_6$ and $[\pm\theta]_{12}$, where θ = 30°, 55° and 85°. A 58 mm long and 13 μm thick, release agent coated aluminum film ,was inserted at the mid-plane of the cylinders during the filament winding process to define a starter delamination crack. After filament winding, the large diameter cylinders were cured at 160°C for three hours. Measured fiber volume fraction was 0.61 for all cylinder lay-ups. Beam fracture specimens of a nominal length of 200 mm and a nominal width of 18 mm were cut from the cylinder wall for the subsequent DCB, ENF and MMB tests schematically illustrated in Fig. 1. The beam axis was parallel to the cylinder axis producing straight beams with a curved cross-section. In order to accommodate the curved cross-section of the beam fracture specimens, contoured aluminum loading tabs were fitted to the DCB and MMB specimens and contoured loading pins and supports were attached to the ENF and MMB test fixtures. Experimental details are provided in Refs [4-6].

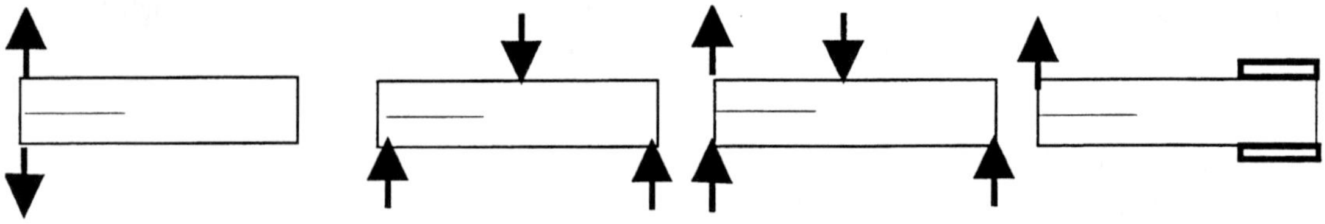

DCB Mode I ENF Mode II MMB (mode I + II) ADCB (Mode I + II)

Figure 1. Configurations of delamination specimens

Glass/epoxy cylinders of lay-up $[\pm 55]_6$ and 110 mm length were manufactured for the external pressure tests using the filament winding process. The internal diameter was 55 mm, the wall thickness 6.5 mm, and fiber volume fraction was 0.68. E-glass fibers in these smaller diameter cylinders were impregnated with the same epoxy resin as the larger diameter cylinders.. The cylinders were cured at 125°C for seven hours. In order to simulate fabrication defects that may arise in thick cylinders due to exothermic heating, 50 mm square aluminum foil layers of 13 μm thickness, coated with release agent on both sides, were introduced at different thickness locations during filament winding. The wall consisted of 12 layers and defects were placed between the 3rd and 4th (referred to as 1/4 thickness), 6th and 7th (mid-thickness), and 9th and 10th layers (3/4 thickness which is closest to the inner wall).

Delamination Fracture Testing

DCB, ENF and MMB specimens were tested, with two to four replicate specimens for each lay-up. All fracture tests were performed from the end of the (straight) insert without pre-cracking. Specimens obtained from unidirectional plates were DCB (initial crack length 50mm), ENF (crack length 25 mm), and ADCB [7] (DCB asymmetrically loaded by clamping the uncracked end of the specimen, free length 110 mm, initial crack length 50 mm and loading only one arm in tension to give a ratio G_I/G_{II} of 4/3). Five or six replicates were used for each test. The specimens were loaded in displacement-controlled testing frames. A slow cross-head speed (between 0.5 and 1 mm/min) allowed for visual monitoring of crack propagation. Displacement of the cross-head was measured with a LVDT, and a real-time load (P) versus displacement (δ) response was recorded. The onset of crack growth from the starter insert was determined by inspection of the specimen edge with a travelling microscope during the test and observation of the P-δ response on the chart recorder. Fracture testing of the curved DCB specimens employed the incremental method proposed by Wilkins et al. [8].

Data reduction for the flat specimens was performed using the experimental compliance calibration for the DCB specimen [9,10]; beam theory for the ENF specimen [11,12] and large displacement corrected beam theory for the ADCB specimen [13]. Energy release rate expressions were derived for the DCB, ENF and MMB cylinder beam specimens to allow for fracture toughness determination. [4-6]. Such analysis incorporates the effect of asymmetry of the delaminated region due to lay-up and thickness, and elastic foundation effects for the finite beam geometries. The experimental compliance calibration method [9,10] was also used to determine fracture toughness of the DCB and ENF specimens. Compliance calibration of the MMB test was not possible and beam theory was used to reduce fracture toughness.

Impact Testing of Cylinders

Impact damage was introduced using steel hemispherical impacters of 34 mm tip diameter and 400 and 600g mass, dropped in the radial direction from 2 meters height at mid-length of horizontally arranged cylinders placed in a cradle. The corresponding impact energies are 8 and 12 J, respectively.

External pressure tests on cylinders

A series of external pressure tests was performed on the cylinders to examine the influence of different types of delamination. The cylinders were placed between two thick steel end blocks and loaded to failure in IFREMER's 24 kbar pressure tank. The length was chosen to be sufficiently short (110mm) to minimize the possibility for global buckling of the cylinders. The principal stresses in a pressurized cylinder are axial and hoop stresses. The magnitude of the hoop stress is twice that of the axial stress [14]. The apparent strength expressed as hoop stress at implosion is calculated using the expression for thin wall pressurized cylinders [14]:

$$\sigma_{\theta\theta} = pR/h \qquad (1)$$

where p is the pressure at implosion, R is the mean radius and h is the wall thickness.

In order to examine the loading of the inner sublaminate of the delaminated region a hoop oriented strain gage was bonded at mid-section to the inner wall. A reference gage was bonded to the diametrically opposite inner surface of the cylinder.

RESULTS AND DISCUSSION

Delamination Fracture Test Results

The composite cylinder DCB specimens displayed initially stable P-δ response with crack propagation loads increasing with increased crack length, corresponding to "R-curve behavior". Interlaced fiber bundles intersecting the crack path arrested crack growth, and subsequent crack growth was observed at another interface. In addition to the interlaced fibers, it was found that pull-out of bridged fibers contributed to the fracture resistance and R-curve behavior. The ENF and MMB specimen also displayed crack jumping and formation of secondary cracks, although their inherent instability made such observations more difficult.

Figure 2 shows fracture toughness, G_c for the cylinder laminates plotted versus the mode II fraction, G_{II}/G. Also included in the graph are the G_c (NL) data for the unidirectional glass/epoxy composite. The initiation toughness increases with increasing mode II fraction (G_{II}/G) and angle θ at the ±θ interface.

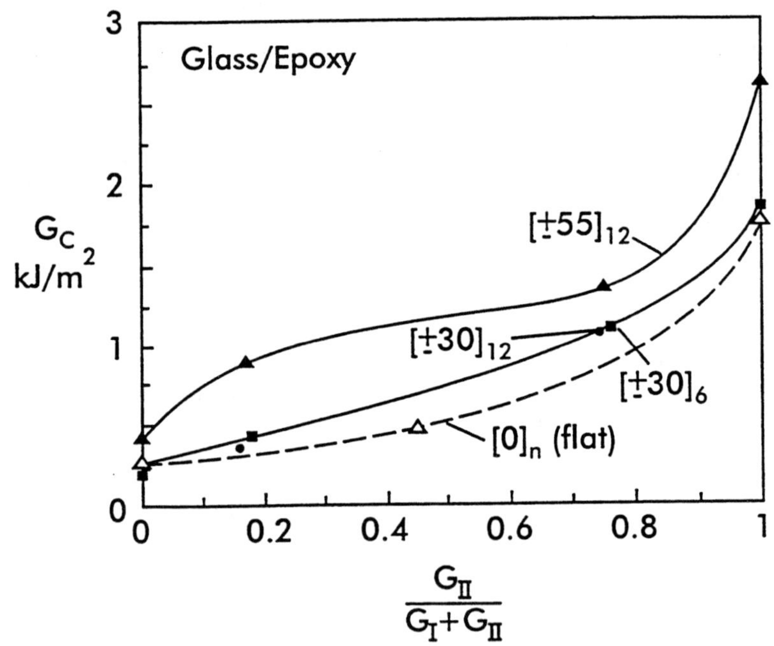

Figure 2. Fracture toughness, G_c, versus mode II fraction (G_{II} / G).

The implications of the toughness trends shown in Fig. 1 for the integrity of a cylinder containing a delamination would require extensive analysis. Although such analysis is not currently available, the work of Kardomateas and Chung [15] shows that a necessary condition for further growth of the delamination is buckling of the sublaminate/s of the delaminated region. Once the delamination buckles, it would tend to propagate in a direction governed by the local balance between available strain energy release rate and fracture resistance at the given mode mixity (Fig. 2).

For the $[\pm 55]_6$ cylinders considered, propagation along the cylinder axis i.e. between a ±55° interface, would be associated with greater resistance than in the hoop direction, where the delamination would propagate along a ±35° interface. It is also expected that the larger stiffness of the wall laminate in the hoop direction than in the axial direction would promote mode I dominated loading of the crack front as discussed by Whitcomb [16]. This would further contribute to extend the delamination in the hoop direction since the toughness is decreasing with increased mode I fraction, see Fig. 2.

External Pressure Test Results

Figure 3 shows the hoop strength for cylinders without and with delamination.

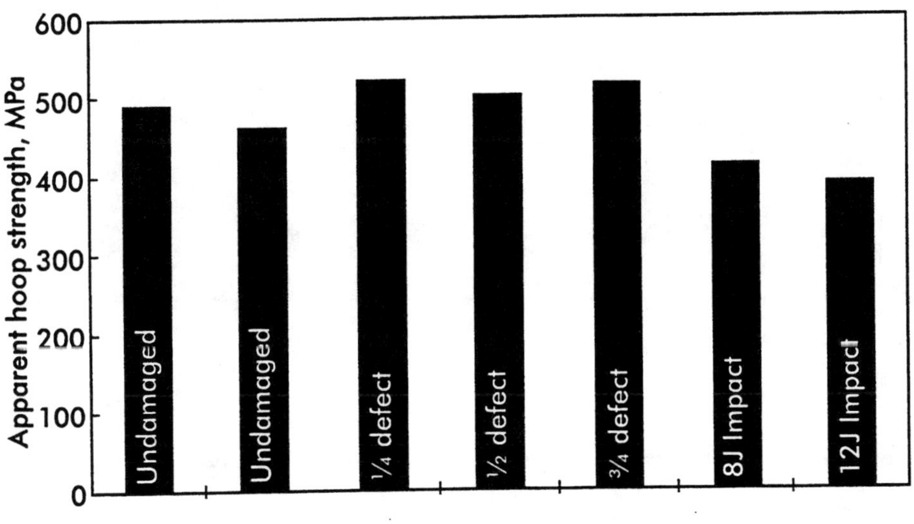

Figure 3. Hoop stress at failure for externally pressurized $[\pm 55]_6$ glass/epoxy cylinders.

The presence of a single delaminations has no significant influence on the implosion pressure. At the start of loading, the strain gages bonded to the sublaminate of the delaminated region responded differently than those on the opposite wall (without defects), Fig. 4. The gages over the defects did not react to the pressure at first. Thus in the case of the ¾ thickness defect, a pressure of 75 bars was required before the inner sublaminate of the delaminated region of the cylinder experienced compression strain, while when the defect is at mid-wall thickness the load is transferred to the delaminated inner wall, at 50 bar pressure. As pressure is increased, the outer part of the cylinder wall then presses against the inner delaminated part, and then there is no further influence of the delamination, as indicated by the similar linear strain responses with and without defects. The responses of the gages on the wall diametrically opposite to the implanted defect, and of both those on the ¼ depth defect cylinder (not shown in Fig. 4), superpose perfectly on the strain response for a cylinder without defects.

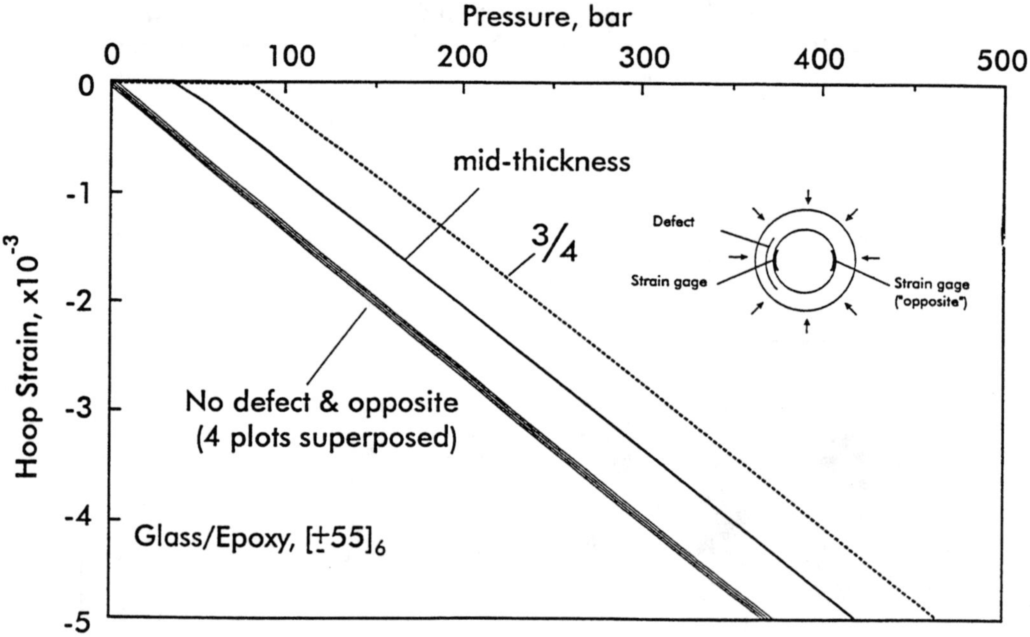

Figure 4. Hoop strain gage responses for cylinders with and without delaminations.

For cylinders impacted at 8 and 12 J, sectioning and examination of the impacted region revealed that multiple delaminations occurred through the wall thickness. The summed area of the delaminated surfaces after an 8J impact was found to be very similar to the area of the artificial defects (2500 mm^2). When the impacted cylinders were pressure tested, local failure occurred, and there was a 15 to 20% reduction in failure pressure, These results suggest that for the [±55°]$_6$ glass/epoxy cylinders examined here, impact damage is likely to be more harmful in service than fabrication defects. However, this may not be the case for other lay-ups and further tests are underway to examine these parameters.

CONCLUSIONS

An experimental study on glass/epoxy angle-ply laminate specimens machined from composite cylinders was conducted over a wide range of mode ratios. Specifically, $[\pm\theta]_6$ and $[\pm\theta]_{12}$ laminates with mid-surface delaminations were considered, where $\theta = 30°$, $55°$ and $85°$. The initiation value of mixed mode fracture toughness, G_c increased with increased mode II fraction and ply angle. Cylinders loaded under external pressure were insensitive to single artificial delaminations, but multiple delaminations of similar total area produced by impact caused reductions in implosion pressure. The insensitive of the failure pressure to a single delamination is attributed to the absence of buckling of the delaminated sublaminates before the cylinder wall collapsed. The impacted cylinders contained multiple delaminations which caused local reduction in the compressive local capability and reduction in failure pressure.

ACKNOWLEDGEMENTS

This research has been conducted within a cooperative program between National French Oceanographic Research Organization, IFREMER, and the Department of Mechanical Engineering at Florida Atlantic University (FAU). The participation of FAU in this program is sponsored by the Office of Naval Research (ONR) with Dr. Yapa D. S. Rajapakse as the program monitor. Technical support in the pressure test programme from members of the Qualification and Testing service at IFREMER, P. Warnier, E. Person A. Deuff and J.J. Le Roy is gratefully acknowledged. Thanks are also due to Mrs. Xiaoming Li and Mr. Gilmer Viana Camps for their assistance with the preparation of this manuscript.

REFERENCES

1. Davies, P., Blackman, B.R.K. and Brunner, A.J., "Standard Test Methods for Delamination Resistance of Composite Materials: Current Status", Appl. Compos. Mater., vol. 5, 1998, pp. 345-364.

2. Riddle R.A. and Beckwith S.W., "Development of Test and Analysis Methods for Thick-Wall Graphite/Epoxy Filament Wound Composite Materials Fracture Toughness", ASTM STP 893, 1986, pp. 64-83.

3. Davies P. and Rannou F., "The Effect of Defects in Tubes", Appl. Compos. Mater., vol.1, 1995, pp. 333-349.

4. Ozdil, F. and Carlsson, L.A., "Characterization of Mode I Delamination Growth in Glass/Epoxy Composite Cylinders", submitted to J. Compos. Mater., 1998.

5. Ozdil, F. and Carlsson, L.A. and Li, X., "Characterization of Mode II Delamination Growth in Glass/Epoxy Composite Cylinders", to appear in J. Compos. Mater., 1999.

6. Ozdil, F. and Carlsson, L.A., "Characterization of Mixed Mode Delamination Growth in Glass/Epoxy Composite Cylinders", submitted to J. Compos. Mater., 1999.

7. Vanderkley P.S., "Mode I - Mode II Delamination Fracture Toughness of a Unidirectional Graphite/Epoxy Composite", MS Thesis, Texas A&M University, December 1981.

8. Wilkins, D.J. Eisenmann, J.R., Camin, R.A., Margolis, W.S. and Benson, R.A., "Characterizing Delamination Growth in Graphite/Epoxy", ASTM STP 775, 1982, pp. 168-183.

9. Berry J.P, "Determination of Fracture Surface Energies by the Cleavage Technique", J. Appl. Phys., vol. 34, 1, 1962 pp. 62-68.

10. Ewalds, H.L. and Wanhill, R.J.H., "Fracture Mechanics", Edward Arnold, London, 1989.

11. Russell A.J. and Street K.N. "Factors Affecting the Interlaminar Fracture Energy of Graphite/Epoxy Laminates", Proc. ICCM-4, 1982, pp. 279-286.

12. Carlsson, L.A, Gillespie, Jr., J.W. and Pipes, R.B., "On the Analysis and Design of the End-Notched Flexure (ENF) Specimen for Mode III Testing", J. Compos. Mater., vol. 20, 1986, pp. 594-604.

13. Wang Y. and Williams J.G. "Corrections for Mode II Fracture Toughness Specimens of Composite Materials", Compos. Sci. Tech., vol. 43, 1992, pp. 251-256.

14. Gere, J.M. and Timoshenko, S.P., "Mechanics of Materials" (4th ed.), 1997, PWS, Boston, MA.

15. Whitcomb, J.D., "Parametric Analytical Study of Instability-Related Delamination Growth", Compos. Sci. Tech.., vol. 25, 1986, pp. 19-48.

16. Kardomateas, G.A. and Chung, C.B., "Thin Film Modeling of Delamination Buckling in Pressure Loaded Laminated Cylindrical Shells", AIAA J. Vol. 30 1992, pp. 2119-2123.

AMD-Vol. 235, Thick Composites for Load Bearing Structures
ASME 1999

HIGH STRAIN RATE RESPONSE OF RESIN INFUSION MOLDED SANDWICH COMPOSITES

Hassan Mahfuz[1], Wahid Al Mamun, Hisham Mohamed, Uday Vaidya, Anwarul Haque and Shaik Jeelani
Tuskegee University's Center for Advanced Materials
Tuskegee, AL 36088.

ABSTRACT

Foam core sandwich composites have been tested under high strain rate (HSR) loading in the thickness direction. The regular Split Hopkinson Pressure Bar (SHPB) has been modified by replacing the steel transmitter bar by a polycarbonate bar. This modification resulted in stronger signals from the transmitter bar, which would otherwise be very feeble especially when testing soft materials. New sets of mathematical formulations have been derived to account for the impedance mismatch between the incidence and transmitter bars. The modified equations are first verified with a known material and then used for sandwich composites. Three types of core with various densities have been tested under compression at strain rates ranging from quasi-static to 1000 S^{-1}. The compressive failure stress has been observed to be directly proportional to the core density, as well as to the strain rate. The strain rate sensitivity was moderate, and the sandwich composites mostly failed by the collapse of the foam-cell. Delamination did not play a major role in the failure process. Details of the mathematical derivations and the analysis of the HSR behavior are presented in this paper.

INTRODUCTION

Our continuing quest for lighter and stronger materials has almost forced us to lean towards composites and sandwich constructions for marine structures. The current thrust of the three major DoD agencies including Navy, Air Force and Army, have identified affordable light structures as their focal points of interests as we move to the next millennium [1-4]. To Navy, Sandwich constructions become especially significant because of three reasons; (1) very low hull weight, (2) low signatures, and (3) high energy absorption during mine blast or ballistic impact. Sandwich constructions provide unique advantages over conventional metals and alloys on all of the three counts. A sandwich construction stiffens a structure without materially increasing its weight. The deformable viscoelastic core also plays an important role in enhancing the energy absorption capability during an impact. It internally dampens the impact energy as the stress waves propagate through the core. It is the intent of this study to understand the response of the sandwich composites under such stress wave loading.

High strain rate characterization of composites has been a subject of great interest among researchers primarily because of the need to understand their behavior in loading situations where they are subjected to a wide range of strain rates [5-6]]. The strain rate sensitivity of the material is something that the designers have to know before designing a structure for dynamic loading situation. In general the subject of high strain rate loading on composites is complex. When this composite is bonded to viscoelastic core to form into a sandwich construction, the analysis becomes even more challenging.

Typically, a Split Hopkinson Pressure Bar (SHPB) is used to acquire the response of materials under high strain rate loading [7-8]. The technique has been extended to composite materials without much of a change except that one has to be careful with the impedance match between the bars and the specimen-material [9-11]. While there are issues still to be resolved, as to how the constituent matrix and fiber will affect the overall composite response [12], SHPB is a popular technique to characterize composite specimens. In case of sandwich composites, the scenario is however, quite different. Due to the presence of the soft-core material the magnitude of strain transmitted to the transmission bar will be significantly small. In a regular SHPB set up with steel or aluminum bars, one cannot obtain accurate strain or stress response with such feeble signal. Some modification of SHPB system to acquire the response of sandwich specimens is therefore, in order.

One idea is to use low impedance material for both incidence and transmission bars so that one can obtain a measurable pulse at the transmission end. A case in point is the use of viscoelastic bars, which has been tried by many researchers [13-16]. Disadvantage with viscoelastic bars is that because of the wave attenuation and dispersion it requires idealized assumptions, which are not necessarily true [17]. Furthermore, the dynamic mechanical properties are also susceptible to environmental effects such as temperature, moisture level and aging factors [17, 18]. Due to these limitations use of viscoelastic bars has not been pursued extensively. Alternately, attempts have been made to use aluminum alloy bars with smaller and hollow cross-sectional area transmitter bar [17]. The goal was to reduce the

[1] Author to whom correspondence should be addressed, ememah@tusk.edu

ratio of the cross-sectional areas between the bar and the specimen such that the transmitted strain can be increased. The concept is good from a theoretical standpoint, but in practice the hollow transmitter bar requires a fitted aluminum cap at its end to hold the specimen. This introduces impedance mismatch and disturbs the stress wave propagation. Although a pulse-shaper is used to filter out the end-cap influence, the question of impedance mismatch still remains.

In the present investigation, a novel approach is adopted to test sandwich composites using a SHPB. Since only the transmitter signal is weak, a polycarbonate bar is used to replace the transmitter bar as shown in Fig. 1. Because of its low impedance, the polycarbonate bar boosts the transmitted signal to an extent that the regular SHPB data acquisition system can capture the transmitted pulse even for a very soft material. The low impedance of polycarbonate is due to its low modulus and density. Use of polycarbonate bar introduces a different set of problems since now the two bar materials are different. A substantial modification of the SHPB equations is therefore, necessary to utilize this system for strain measurements. Detailed description of the mathematical modifications is presented in this paper. Several foam-core sandwich specimens have also been tested using the polycarbonate transmitter bar, and the analysis of their stress and strain behavior is included.

MATHEMATICAL MODIFICATIONS

In a regular Split Hopkinson Pressure Bar the nominal strain ε (t) in the specimen is calculated as [7]:

$$\dot{\varepsilon}(t) = -\frac{2C_0}{L}\varepsilon_r(t) \tag{1}$$

Where L is the original length of the specimen, ε_r (t) is the time-resolved strain of the reflected pulse in the incident bar, and C_0 is the wave velocity through the bar material. Integration of eqn. (1) with respect to time gives the time-resolved axial strain of the specimen. The nominal axial stress, σ, in the specimen is determined using the equation

$$\sigma(t) = \frac{A_0}{A_s}E_0\varepsilon_t(t) \tag{2}$$

Where A_s is the cross-sectional area of the specimen, ε_t is the time-resolved axial strain in the transmission bar of cross-sectional area A_0, and E_0 is the Young's modulus of the bar material. The foregoing calculations are based on the assumptions that the specimen undergoes homogeneous deformation and that both the incidence and transmitter bar are made of the same material, and are of same cross-sectional area.

If now a polycarbonate bar replaces the transmitter bar of maraging steel, equation (1) and (2) will no longer be applicable. Materials of the incidence and transmission bars are now different. This difference in materials can be accounted for in the following way.

Let u_1 and u_2, be the specimen-bar interface velocities at the incidence and transmission faces respectively. Now if we assume that the particle velocity is equal to the interface velocity, and that part of the incoming strain is reflected at the incidence end, then it can be easily shown that these time-resolved displacements are:

$$u_1(t) = C_i\int_0^t[\varepsilon_i(t) - \varepsilon_r(t)]dt \qquad \text{at } t>0 \tag{3}$$

and $$u_2(t) = C_t\int_0^t\varepsilon_t(t)dt \tag{4}$$

where $\varepsilon_i(t)$ is the incident strain pulse measured by a strain gauge mounted axially on the incident bar surface, C_i and C_t are the incident bar and transmission bar wave velocities respectively. The axial engineering strain of the specimen is therefore,

$$\dot{\varepsilon}(t) = \frac{u_1(t) - u_2(t)}{L} \tag{5}$$

$$\therefore \dot{\varepsilon}(t) = \frac{1}{L} \int \{C_i[\varepsilon_i(t) - \varepsilon_r(t)] - C_t\varepsilon_t(t)\}dt$$

When the specimen is in stress equilibrium, the axial forces acting at the two interfaces are equal, and hence,

$$E_i(\varepsilon_i + \varepsilon_r)A_0 = E_t\varepsilon_t A_0$$

or $$\varepsilon_t = \frac{E_i}{E_t}(\varepsilon_i + \varepsilon_r) \tag{6}$$

where E_i and E_t are the Young's modulus of the materials of the incident and transmission bars respectively. Substitution of eqn (6) into eqn (5) gives

$$\varepsilon(t) = \frac{1}{L} \int_0^t \left\{ C_i[\varepsilon_i(t) - \varepsilon_r(t)] - C_t \frac{E_i}{E_t}(\varepsilon_i + \varepsilon_r) \right\} dt$$

or, $$\varepsilon(t) = \frac{C_i}{L}\left[\left(1 - \frac{C_t}{C_i}\frac{E_i}{E_t}\right) \int_0^t \varepsilon_i(t)dt - \left(1 + \frac{C_t}{C_i}\frac{E_i}{E_t}\right) \int_0^t \varepsilon_r(t)dt \right]$$

$$\therefore \varepsilon(t) = \frac{C_i}{L}\left[(1-K)\int_0^t \varepsilon_i(t)dt - (1+K)\int_0^t \varepsilon_r(t)dt \right] \tag{7}$$

where, $K = \frac{C_t}{C_i}\frac{E_i}{E_t}$

Eqn (7) can now be used to calculate nominal axial strain of the specimen from the measured incident and reflected strain pulses using the modified SHPB setup.

In order to obtain the average stress in the specimen after the equilibrium is reached,

$$\sigma = \frac{P_1(t) + P_2(t)}{2A_0} \tag{8}$$

Where $P_1(t)$ and $P_2(t)$ are the forces acting at interfaces 1 and 2 as defined previously. Since,

$$P_1(t) = A_0 E_i(\varepsilon_i + \varepsilon_r)$$

and $$P_2(t) = A_0 \varepsilon_t(t) \tag{9}$$

Substituting eqns (9) and (6) into (8), one can show that

$$\sigma = \frac{A_0}{A_s} E_t \varepsilon_t(t) \tag{10}$$

If now the strain pulses, namely, $\varepsilon_i(t)$, ε_r (t), and ε_t (t) are measurable, strain and stress on the sample can be determined from equations (7) and (10) respectively.

MATERIALS FABRICATION

The Resin Infusion Molding (RIM) process was employed to fabricate the sandwich panels. Since both the top and bottom face sheets had to be infused simultaneously, a co-injection resin infusion process was developed to process the sandwich composites. A schematic of the co-injection process is shown in Fig. 2. Dry fabric preforms with required orientations are first laid out on the top of a flat aluminum tool. The fabric used in this investigation was twill weave S2-Glass fibers. Six layers of fabrics were used for each face sheets. The core was then placed on the top of the bottom face sheet fabrics, and upon which the preforms for the top face sheets were stacked. Three core materials were used during the fabrication. The foam types were; Klegcell R-75, H-100 and H-130 manufactured by Divinycell International Inc. The Core material was semi-rigid PVC core specially designed for production of FRP sandwich structures. After stacking, the assembly was vacuum bagged, and infusion lines were installed. Two resin infusion lines and one vacuum line as shown in Fig. 2 were used. Before infusion the system was debulked for several hours. Vinyl ester (Dow Derakane 411-350) resin manufactured by Dow Corning was used in this study. After the resin infusion, the vacuum was kept on until the complete cure took place. No additional adhesives were used for the skin-core bonding, since it developed naturally during the cure process. Several panels were fabricated in this manner, and were used for machining coupons for the SHPB tests. The thickness of each face sheet was around 0.15 inch, while that of core was about 1.0 inch.

RESULTS AND DISCUSSION

Effect of Viscoelastic Properties

Since polycarbonate bar has viscoelastic properties, it is expected that the stress wave propagating through this bar will be viscoelastic in nature rather than elastic as it is in case of aluminum or steel bars [20, 14]. This raises the question whether a viscoelastic wave analysis is necessary to increase the accuracy of the measurements. In the present case, only the transmitter bar was made from polycarbonate, while the incident bar remained elastic (steel). It is therefore, expected that attenuation and distortion of the strain pulses, namely, incidence and reflected, will be minimal, since both of these pulses are being measured in the elastic bar. Again, these two pulses are being used in eqn (7) to compute the sample strain suggesting that any error from the viscoelastic properties of the transmitter bar would be insignificant. However, one should be careful in using eqn (7) since it requires C_t and E_t, which correspond to the polycarbonate bar. Determination of these parameters is given in a following section.

Before we discuss the determination of C_t and E_t, it is important that we do investigate the shape of the pulse, and sample strain measured by steel-polycarbonate and steel-steel systems separately. This was done by testing few cylindrical specimens of Aluminum in the Hopkinson bar set up. In one set both the bars were of steel while in the other set the steel transmitter bar was replaced by a polycarbonate bar. The idea was to get the response of a known material by both methods and do a comparison. Figure 3a shows the strain pulses measured from the steel-steel system. As expected, the incidence and reflected pulses are identical without any attenuation or dispersion. The corresponding transmitter pulse also shows the same behavior as shown in Fig. 3b. Now if we look at Fig. 4a, which shows the strain pulses measured from the steel-polycarbonate system, the incidence and the reflected pulses are similar to what we observed in Fig. 3a. The reason for similarity is due to the use of steel incidence bars in both cases. However, the transmitted pulses as seen in Fig. 4b are somewhat different from what was seen in Fig. 3b. The first and second peaks are both incoming transmitted pulses, while the third pulse is the reflected one from the free end of polycarbonate bar. Since the wave speed is low in polycarbonate, it took longer time for the reflected pulse to arrive at the gage location, and that happened after the second pulse got there. We observe slight attenuation and distortion of the waves with the passage of time. It is also noticed that the amplitude of the transmitted pulse has increased many folds, almost by a factor of 10. In Fig. 3b, the amplitude of the transmitted pulse is around 130 mV, while it is approximately 1300 mV in Fig. 4b. This clearly justifies that the use of polycarbonate bar will be most appropriate for testing sandwich composites. As a final check, sample strains on aluminum specimens were calculated using eqn (7). These strains with respect to time are shown in Fig. 5. The curves of Fig. 5 are not identical, but good similarity between the two can be observed. Maximum strain measured by both systems varies between 2.7% and 2.9%. What this establishes is that one can use a steel-polycarbonate system even for aluminum without introducing significant error.

Determination of C_t and E_t for the Transmitter Bar

It was essential that the wave velocity, C_t and modulus, E_t of the transmitter bar be determined accurately especially within the expected strain range of the test. Strain gages were bonded onto the polycarbonate bar at a known distance from free end, and the bar was then mounted as transmitter bar in the SHPB setup in direct contact with the steel incidence bar. Strain signals were generated by impacting through a striker bar, and then the signals were recorded in the data acquisition system. The pulse recorded on the polycarbonate bar is shown in Fig. 6. The first pulse is the incoming pulse while the second one is the reflected pulse as the wave is reflected from the free end. The reason that pulse shapes are different from that of Fig. 4 is that there is no specimen between the incidence and transmitter bar at this time. The time difference between the two events is approximately 1000 μsec as seen in Fig. 6. Since now the distance between the location of the gage and the free end of the bar is known, one can easily determine the wave velocity, C_t. Once C_t is known, E_t can be calculated by multiplying the square of C_t with the density of the bar material. The density of polycarbonate as obtained from the manufacturer was 0.000112 slug/in^3. Several tests were conducted in this manner at various strain rates ranging from 100 - 500 S^{-1}. There was slight variation in the values of E_t, but they all remained within 3.5 - 4.0 X10^5 psi. An average value of 3.73 X10^5 psi was used for our calculations. The quasi-static value for E_t as supplied by the manufacturer was 3.1 X10^5 psi, which is reasonably close to our dynamically determined value. The corresponding wave speed through the polycarbonate bar was calculated to be 4806 ft/sec.

Sandwich Response

Sandwich specimens were made from three different types of foam namely, R-75, H-100, and H-130. The difference between them was their densities indicated by the numbers attached to each category. For example, H-100 represented a density of 100 Kg/m^3, while H-130 meant a density of 130 Kg/m^3. Strain versus time response of R-75 sandwich at three different strain rates is shown in Fig. 7. It is observed that the peak value of strains increases with increasing strain rates. Peak sample strains range from 2% to 8% as strain rates vary from 380 S^{-1} to 973 S^{-1}. The time to reach the peak strain is however, seen to be same for all strain rates, around 95 μsec. This allows the specimen to be strained more in case of higher strain rates. The flat nature of these curves is somewhat different than what is usually observed with metals and composites. This is due to the presence of the foam, which is mostly dominating the performance of the sandwich at high strain rates. Similar pattern was observed with H-100 and H-130 type sandwiches.

After analyzing the strain response, it was important that we look at the stress-strain behavior to shed some light into the failure modes of the sandwich composites. At first, a few samples were tested quasi-statically in a regular MTS machine under displacement control, and the stress strain curves were generated as shown in Fig. 8. The strain rate for the quasi-static test was approximately 0.001 S^{-1}. It is observed in Fig. 8 that with the increase in density, there is a sharp increase in failure stress. This increase is more prominent between R-75 and H-100 rather than between H-100 and H-130 suggesting that there is no proportionate increase in strength with density. The failure stress for R-75, H-100 and H-130 are respectively 120 psi, 250 psi and 350 psi which are very well comparable to 160 psi, 247 psi and 363 psi given by the manufacturer for their foams. It is also noticed in Fig. 8 that failure strains at quasi-static loading ranged from 1.5% to 3.0%. A quick calculation of the initial stiffness gives the modulus of R75 to be around 8,000 psi, which is slightly higher than 5,503 psi given by the manufacturer. This increase in quasi-static modulus is due to the presence of composite face sheets in the sandwich. Each curve of Fig. 8 rises uniformly with load until the peak is reached, and it is believed that at that point the foam cells have collapsed completely. It is no longer capable of sustaining any load, and deformation increases uncontrollably, at which point the machine actually stops. There are neither load drops nor any changes in the slopes of these curves before they reached the peak suggesting that no delamination occurred during the quasi-static loading.

At higher strain rates the stress-strain behavior is quite different as seen in Fig. 9. Failure strains now vary from 3.5% to around 8%, the lowest being for the H-130 sandwich. This is just opposite to what we observed in Fig. 8, where failure strains were higher for higher densities. The reason for such reversal at high strain rate is due to the fact that higher density foam can absorb less energy than that of low-density foam. Larger number of empty cells in low-density foam would allow more and quicker dissipation of impact energy. This is also reflected in the slope of H-100 and R-75 curves. Maximum stress sustained by each category follows the same trend as it was observed in Fig. 8. It is also noticed that there is a moderate increase in failure stress as one moves from quasi-static to high strain rate loading.

Tests were conducted at four different strain rates for each category of sandwich composites, and their variation with strain rates are shown in Fig. 10. The strain rates varied from 0.001 S^{-1} to 1000 S^{-1}. As expected, higher density foam demonstrated higher failure stress throughout the entire range of strain rates. The slope of the H-130 curve is higher than the other two curves, indicating that as the density increases, so does the strain rate sensitivity of the sandwich. A discrepancy seen in Fig 10 for the H-130 curve is most likely due to the statistical dispersion in the test data taken at two close-range strain rates. It is also noticed that the increase in stress is almost linear for the three cases. The stresses increase by factors of 1.25, 2.0 and 2.38 for R-75, H-100 and H-130 respectively as strain rate changes from quasi-static to 1000 S^{-1}.

SUMMARY

The following can be summarized from the above investigation:

- A novel steel-polycarbonate system has been introduced in a regular SHPB to test sandwich composites at high strain rates.
- New sets of mathematical equations are derived to extract strain and stress responses.

- Viscoelastic nature of the polycarbonate bar has been investigated, and its effect is seen to be minimal on the strain measurements.
- Sandwich composites are moderately sensitive to strain rate, and this sensitivity increases with the density of the core material.
- At both quasi-static and high strain loading, the compressive failure stress of the sandwich is directly related to the core density.
- High strain rate loading does not appear to cause any delamination between the core and the skin. Much of the impact energy seems to be absorbed by the deformation of core alone.

ACKNOWLEDGEMENT

The authors acknowledge with appreciation the support for this research from the Office of Naval Research (ONR) through grant No. N00014-90-J-11995/P00011. Dr. Y. D. S. Rajapakse is the Technical Monitor.

REFERENCES

1. Air Vehicles Directorate Technology Area Plan - FY 98, Internet Release Information, www.wrs.afrl.mil.
2. Surface Ship Combatant Technology, ONR Overview, 4th April 1998, Jim Gagork, ONR Meeting, Biloxi, MS.
3. Composite Armored Vehicle Overview, United Defense Limited Partnership, March 14, 1997, George Thomas.
4. Composite Aircraft Structures, R. Glasco, Elsevier Publishers.
5. Thiruppukuzhi, S.V. and Sun, C.T., "Testing and Modeling High Strain Rate Behavior of Polymeric Composites," Composites Part B: Engineering, v 29, n 5 1998, pp. 535-546.
6. Nicholas, T., "Material Behavior at High Strain Rates," Chapter 8, Impact Dynamics, Zukas, J.A., et al, John Wiley and Sons, New York, NY.
7. Klosky, H.," An Investigation of the Mechanical Properties at Very high Rates of Loading," Proc. R. Soc. London, B62, 676-700 (1949).
8. Lindholm, U.S., "Some Experiments with the Split Hopkinson Pressure Bar," J. mech. Phys. Solids, 12, 317-35 (1964).
9. Harding, J., "Effect of Strain Rate and Specimen Geometry on the Compressive Strength of Woven Glass-Reinforced Epoxy Laminates," Composites, Vol. 24, N 4, 1993, pp. 323-332.
10. Groves, S., Roberto, S., Lyon, R. and Brown, A., "High Strain rate Effects for Composite Materials," Composite Materials Testing and design (11th Volume), ASTM STP 1206, E. T., Camponeschi, Jr. Ed., American Society for Testing and Materials, Philadelphia, 1993, pp. 162-176.
11. Montiel, D. and Williams, C., "A Method for evaluating the High Strain Rate Compressive properties of Composite Materials," Composite Materials: Testing and design (10th volume), ASTM STP 1120, Glenn C. Grimes, Ed., American Society for Testing and Materials, Philadelphia, 1992, pp. 54-65.
12. Mahfuz, H., Mamun, W., Haque, A., Vaidya, U. and Jeelani, S., "Response of Fiber and Matrix in Composites under High Strain Rate Loading," Proceedings of the SEM Annual Conference on Theoretical, Experimental and Computational Mechanics, June 7-9, 1999, Vol. 1, pp. 559-564.
13. Gamby, D. and Chaoufi, J., "Asymptotic Analysis of Wave Propagation in a Finite Viscoplastic Bar," Acta Mechanica, 87 (19191), 163-178.
14. Wang, L., Labibes, K., Azari, Z., and Pluvinage, G., "Generalization of Split Hopkinson Bar Technique to use Viscoelastic Bars," International Journal of Impact Engineering, Vol. 15, n 5, 1994, pp. 669-686.
15. Bragov, A.M. and Lomunov, AK, "Methodological Aspects of Studying Dynamic Material Properties Using the Klosky Method," International Journal of Impact Engineering, Vol. 16, n 2, 1995, pp. 321-330.
16. Zhao, H., Gary G. and Klepaczko, JR, "On the Use of a Viscoelastic Split Hopkinson Pressure Bar, International Journal of Impact Engineering, Vol. 19, n 4, 1997, pp. 319-330.
17. Chen, W., Zhang, B. and Forrestal, M.J., "A Split Hopkinson Bar Technique for Low-Impedance Materials," Experimental Mechanics, June 1999, Vol. 39, n 2, pp. 81-85.
18. Knauss, W.G., Boyce, M., McKenna, G., and Wineman, A., "Non-linear time-dependent Constitution of Engineering Polymers," Report No. 95-10, Institute for Mechanics and Materials, University of California, San Diego, 1995, 1-16.
19. Meyers, MA, Dynamic Behavior of Materials, Chapter 12, John Wiley & Sons, Inc., 1994, New York, NY.
20. Chu, C, Wang, L. and Xu, D., "A non-linear Thermo-viscoelastic Constitutive Equation for Thermoset Plastics at High Strain Rates," Proc. Int. Conf. Nonlinear Mechanics, Oct 28-31, 1985, Shanghai, China, pp. 92.

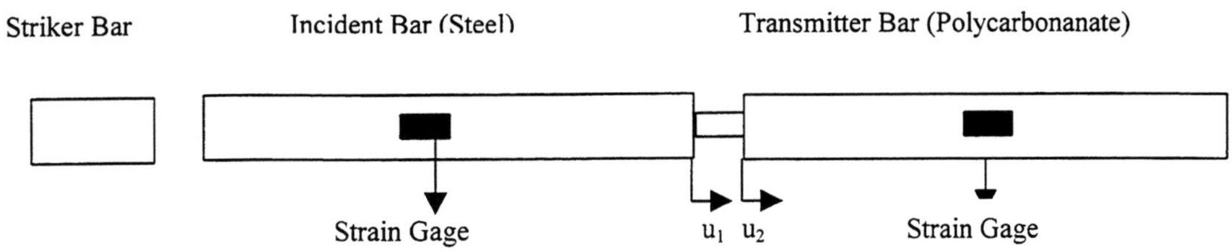

Fig.1: Modified Hopkinson bar

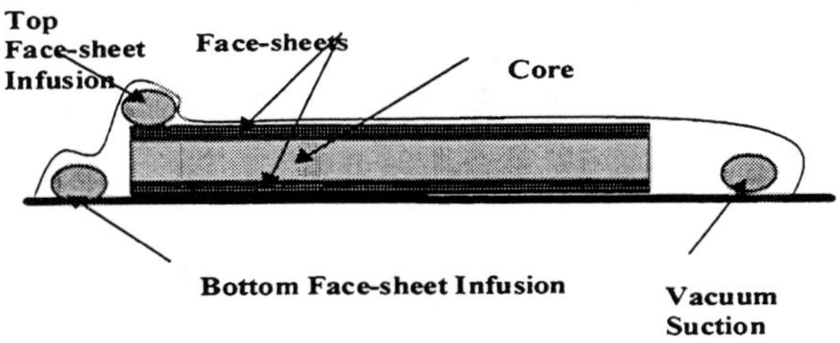

Fig.2: Schematic of the co-injection process for sandwich composites

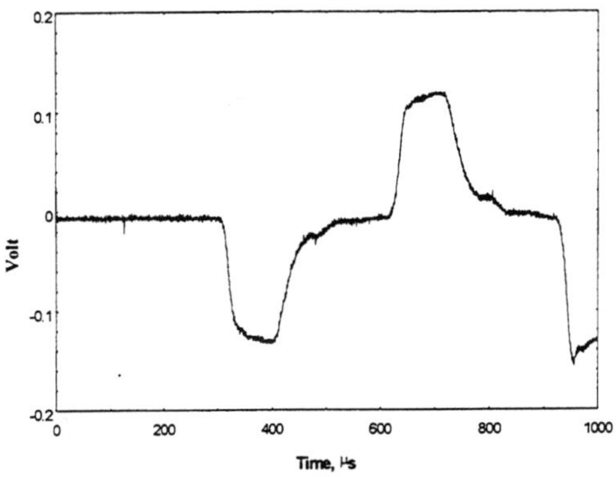

Fig.3b: Transmitter bar response of steel-steel system for Aluminum

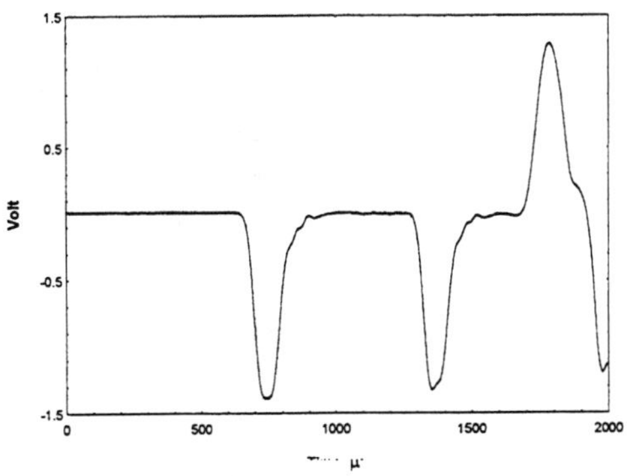

Fig.4b: Transmitter bar response of steel-polycarbonate system for Aluminum

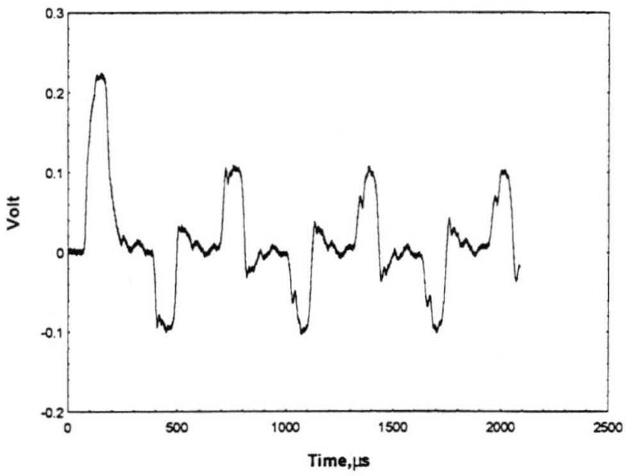

Fig.3a: Incidence bar response of steel-steel system for Aluminum

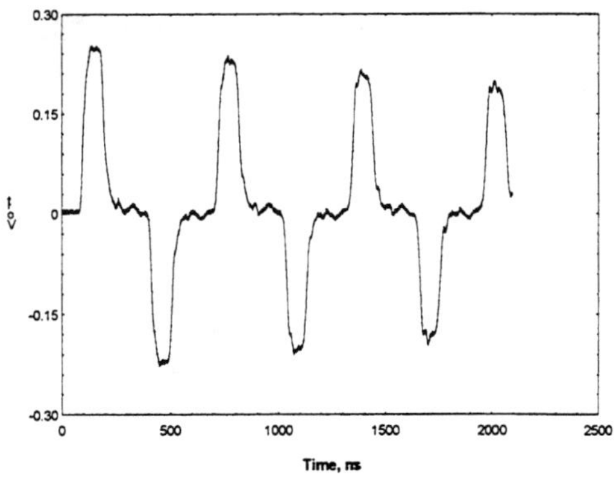

Fig.4a: Incidence bar response of steel–ploycarbonate system for Aluminum

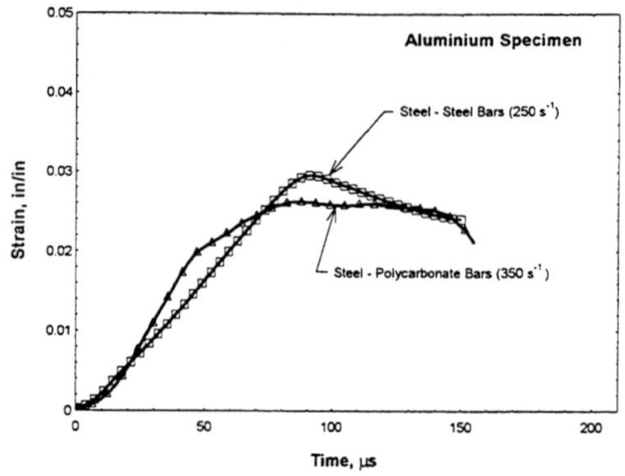

Fig.5: Comparison of strain-time response for Aluminum with two different systems

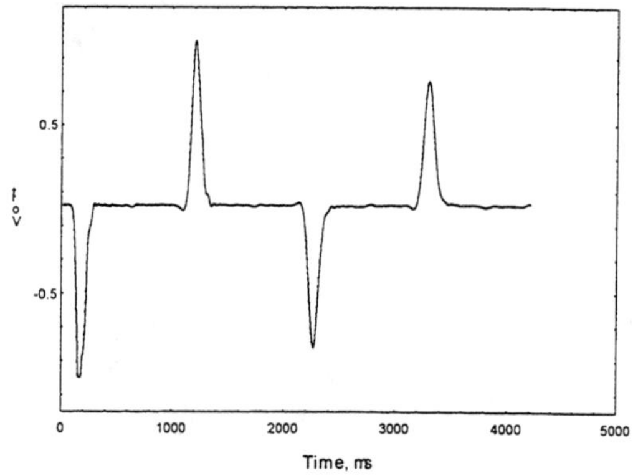

Fig.6: Calibration of Polycarbonate transmitter bar

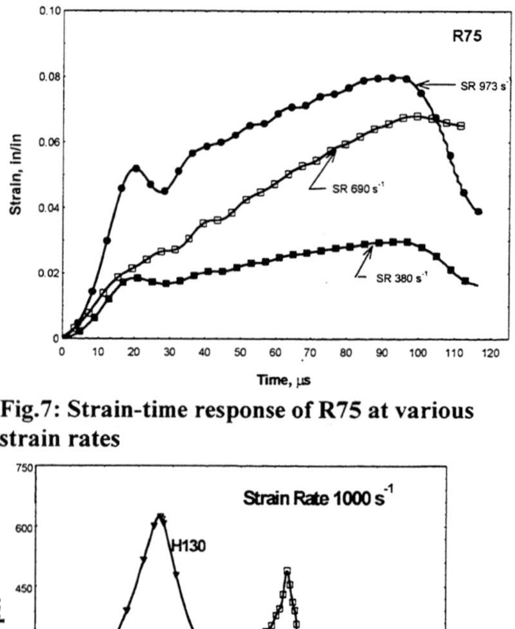

Fig.7: Strain-time response of R75 at various strain rates

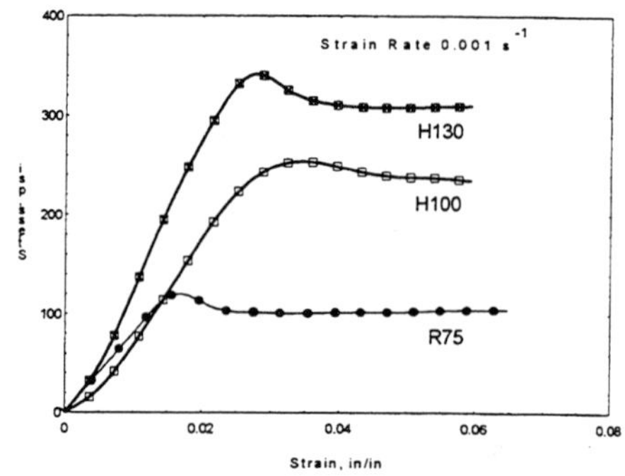

Fig.8: Quasi-static response of three different sandwich composites

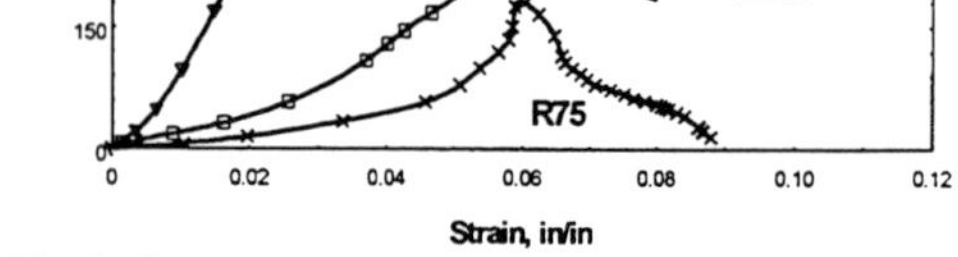

Fig.9: Stress-strain behavior of three sandwich composites at high strain rate

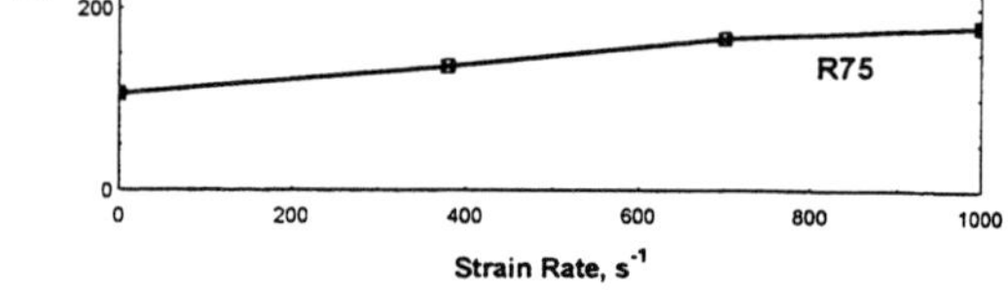

Fig. 10: Failure stress variation with strain rate

AUTHOR INDEX

AMD-Vol. 235
Thick Composites for Load Bearing Structures

Book Number: G01146